T0100183

THE HIDDEN HISTORY OF CODE-BREAKING

THE HIDDEN HISTORY OF CODE-BREAKING

The Secret World of Cyphers, Uncrackable Codes, and Elusive Encryptions

SINCLAIR McKAY

PEGASUS BOOKS
NEW YORK LONDON

THE HIDDEN HISTORY OF CODE-BREAKING

Pegasus Books, Ltd.
148 West 37th Street, 13th Floor
New York, NY 10018

First Pegasus Books cloth edition August 2023

ISBN: 978-1-63936-434-3

10 9 8 7 6 5 4 3 2 1

Printed in the United States of America
Distributed by Simon & Schuster
www.pegasusbooks.com

CONTENTS

INTRODUCTION

Just by the mere act of reading this sentence you have successfully broken a code. All languages are ciphers in a way. Alphabets, grammars and vocabularies – their meanings remain hidden until we can read them. But of course, the purpose of language is clarity of communication and meaning made plain. Whereas the history of codebreaking, in stark contrast, is the story of languages seen through a glass darkly, distorted and obscured, reduced to base constituents of incomprehensible gobbledegook.

For practically as long as there has been writing there have been those – from spies to court officials to mystics – who have sought to disguise their words so that only the initiated could read them. Codes are fundamentally about secrets, and the world has always needed secrets. From the earliest ciphers used by the Ancient Greeks to the quantum-computing wonders of today, the principle has always remained the same. Those wishing to keep their secrets well hidden must find a way to radically disguise and rearrange sentences and letters so that their opponents or rivals cannot detect any method of unscrambling them.

Perhaps the most famous encryption technique of all was the one that was utilised through the German Enigma machine in the Second World War, where electrically generated ciphers were created by a simple-looking portable device of Bakelite and brass with a keyboard and lights. These machines were capable of producing potentially millions upon millions of different code combinations, and the story of how they were secretly defeated – first by Polish mathematicians and then by a brilliant team based at a country house in Buckinghamshire – is rightly celebrated today. But there are so many other instances, throughout history, where codes have proved to be turning points for humanity.

As well as the hieroglyphs found on Ancient Greek and Ancient Egyptian tombs, which remained baffling mysteries for many centuries after those stones were sealed, we find elaborate codes in the Old Testament, ciphers in wide use among Caesar's armies, and even rich and strange carvings on moorland standing stones that conveyed cryptic messages to Vikings. There were codes in the violent courts of the Tudor monarchs and in the conspiracy-filled corners of the Vatican. There were cipher geniuses emerging from the young Islamic civilisation in Babylon, encryption techniques devised amid the art and the splendours of the Renaissance, and codes used as tools in an era of rapacious empire-building that swallowed the world. From Mary, Queen of Scots to Mata Hari, we find an extraordinary underground river of history where crucial moments rested upon secret codes. We also discover how ciphers were used for different, more passionate purposes. Codes of love can be found across continents and times, located in the Kama Sutra or used by the famous eroticist Casanova.

And the story includes many other miracles of Second World War codebreaking performed at Bletchley Park: astounding breakthroughs achieved not by machinery but by the fizzing, dazzling creativity of the minds of the young women and men who were brilliant at maths and linguistics, and super-talented when it came to turning problems on their heads. The unravelling of the Nazi codes was more than just a wonkish intellectual triumph. This particular miracle, performed over and over again, had the most direct effect on all theatres of war and saved literally uncountable lives.

One of the continually marvellous elements of the code-breaking story is just how innocent and simple some of these ingenious encryption techniques were. You will see soon the elegance of the Polybius Square, the beauty of Dr Dee's rotating cipher discs and the baffling wonder of Playfair – a Victorian cube of letters said to be straightforward enough for schoolchildren to master, if not diplomats! Even the Ancient Greek ruse with hexagonal rods and strips of leather with letters written on them is ingenious and offers us a fresh glimpse into lives that are otherwise shrouded in the dust of history books.

But there are stranger depths to be explored as well, with codes that take us into the uncanny realm of mysticism and ciphers that speak of philosophy. What was the lost language of the angels? And who would dare to decode it? What was the truth of the messages that genius inventor Nikola Tesla received from the depths of space? Did encrypted documents discovered within an old French chapel prove an extra dimension to Christianity that sinister forces had spent centuries trying to suppress?

And back on the mortal plain, there were codes in history that spoke of unfathomable courage, such as the special poem ciphers devised for the brilliant female agents of the Special Operations Executive in the Second World War. These are now intensely poignant and they serve as tributes to bravery and sacrifice that very few would have been capable of. There is the story of how a trio of sailors in violent seas dived into a stricken, sinking German submarine in order to capture its codebooks and Enigma machine, an act of nerve that helped turn the Battle of the Atlantic.

There are codes that have transformed the way that we all live. It was almost 200 years ago that a rather talented portrait artist had an idea for sending messages right the way across a continent in the blink of an eye, but which necessitated letters being transformed into encoded sounds. His name was Samuel Morse, and the code to which he gave his name is still in use even in our digital age. There are other sorts of codes too – including those that make up all the elements of our DNA. The unravelling of the genome code in recent years has already proved a revolution in medicine.

Codes, it's clear, are frequently accompanied by wonderful stories and eccentric characters. But it's one thing to learn of the code and another to actually solve it. The best way to celebrate these fantastic decryption achievements is to set a few tests and puzzles of our own along the way. A few will involve actual codebreaking techniques as devised by a variety of figures from around the world, some will be based upon the fearsome challenges that were posed during the world wars, and many others will be tests of mental agility and gymnastics of the type relished by cryptologists. Codebreakers through the ages have been noted for feats of logic and lateral thinking, an ability to hold abstract theories or calculations in their heads as they try different ways to crack problems. The puzzles we have assembled here encompass all the skills and disciplines required for cryptology.

Here, then, is a parallel history of the world and indeed of the written word. Running like hidden networks through all the lands of the world, from the rich red sands of the Silk Road to the glories of Ancient Persia, from the frozen steppes of Russia to the manicured affluence of mid-twentieth-century Washington DC. These stories of those who devised the most fiendish of ciphers, and the stories of the geniuses who used incredible skill to thwart them, run alongside many of the key developments of the last two or even three thousand years. Discover the codes that changed the course of history and the encryptions that changed the nature of civilisations – tales once firmly hidden from view, now brought into the light.

CHAPTER ONE

THE CODES OF THE ANCIENTS

In which centuries-old codes and ciphers are solved in the modern world, forming amazing links between the far-distant dead and the present day . . .

1. THE PALACE OF SECRETS

There was once a code that was a portal through time. It came from an era of classical myth, of heroes, adventures and nightmarish creatures. This mysterious unreadable writing reached through the darkness of the centuries to wholly inhabit the imaginations and intellects of two very different people in the 1940s. One was a female college lecturer working amid the honking horns and heat hazes of New York; the other was a teenage schoolboy in drizzly England. Between them they would decipher a mystery that was thousands of years old.

Amid the bustling streets of 1940s Brooklyn, there was (and is) a tranquil red-brick campus: Brooklyn College. Working there, teaching Latin and Greek, was a pioneering and popular professor called Alice Kober. Her name has, almost accidentally, been erased from the story of one of the greatest codebreaking coups in history. Yet her amazing efforts were the key to understanding what had been one of the most tantalising enigmas: clay tablets inscribed with a language that no one in the modern world could fathom.

This wasn't the sort of code that won wars – rather, it was a puzzle that when solved helped to throw dazzling light upon one of civilisation's greatest myths. The laudatory praise – when it came – would be for a brilliant young British man called Michael Ventris who, after years of obsessive study that started in his schooldays, finally wrenched open the greatest of linguistic challenges. And his honours were well deserved. But this brilliant coup was a shared one.

The myth centred upon the ancient Palace of Knossos, on Crete. The ruins of the site, built around four thousand years ago, had inspired poetic legends for generations, including that of the palace labyrinth, and the fearsome half-man-half-bull Minotaur that was trapped there. In the story, King Minos had the brilliant architect Daedalus design the maze with the creature at the centre of it. The Minotaur was an abomination, the offspring of Minos's wife and Poseidon's snow-white bull. To keep it under control, King Minos offered sacrifices – prisoners from Athens, his less powerful neighbour – every ninth year to satiate the beast's hunger. Theseus, a young prince from Athens, sailed to Crete to kill the Minotaur with the aid of King Minos's daughter Ariadne, who loved him. It was Ariadne who gave Theseus the ball of thread that he could unspool to enable him to find his way to the Minotaur, vanquish it, and find his way out of the impossible labyrinth again. Theseus was successful, but he had coldly used Ariadne, and in one version of the tale, he left her desolate on the island of Naxos on his way back to Athens.

At the beginning of the twentieth century, archaeological investigations at the Palace of Knossos ruins, conducted by Sir Arthur Evans, unearthed examples of ancient writing on clay tablets. These written texts, in different and wholly unknown tongues, were termed Linear A and Linear B. The discovery was at once wonderful and maddening. Here was the potential to go back thousands of years, to understand more about the real lives and civilisation of the Minoans. But the language seemed impossible to decrypt.

Over in America, it was Alice Kober who, in the 1930s and 1940s, became so mesmerised by the possibilities of Linear B that she set about teaching herself a variety of other ancient languages, from Sumerian to Hittite, in order to begin prying open its secrets. This was how she spent all her spare time at home. Strikingly, in keeping careful tabulations of all the symbols and 'letters' of Linear B, Professor Kober was foreshadowing one of the key techniques later to be used by the Bletchley Park operation in the Second World War. She made a meticulous file index of every symbol, every potential term or word, and she analysed the frequency with which each of the symbols turned up and how they related to the frequency of other symbols. Professor Kober used cut-up cigarette cartons as index cards. She seemed wholly undaunted at the prospect of deciphering an entire Bronze Age language.

After years of research, the method began to yield results. She observed that groupings of symbols would start in the same way but end differently, which led her to understand that this was not simply a set of hieroglyphs but an inflected language where the word endings changed according to the context of the sentence.

In the meantime, across the Atlantic, Michael Ventris was also devoting much of his spare time to this ancient enigma. Born in 1922, this young man had once been on a school outing to a museum when, by coincidence, he came across the archaeologist Sir Arthur Evans giving a talk about his discovery of the Linear B tablet. Ventris was immediately grabbed by the idea of unravelling it. Born to multilingual parents, he himself was precociously skilful with languages. Just several years later, when he was only eighteen, he had an article published in an American archaeology magazine, expounding his theory that Linear B was a variation of an ancient Etruscan tongue.

This was no way to earn a living, though, and so young Ventris began training as an architect. As the Second World War broke out, he joined up with the Royal Air Force and served in the hazardous position of navigator on bombing missions flying deep into enemy territory. He married aged twenty and after the war was briefly sent into defeated Germany, partly to help liaise with the Soviets (he spoke both German and Russian). Then came peace, and Ventris continued with architecture; but in all that time, his imagination had remained filled with Linear B and its mysteries (how the otherwise omnipotent recruiters at Bletchley Park had failed to spot this most obvious of codebreaking talents is not clear).

By 1948, his painstaking work (like Professor Kober, Ventris studied a range of dead languages) had attracted serious attention. Following the death of Sir Arthur Evans, the custodian of Linear B was now the academic Sir John Myres, who decided he wanted to combine Ventris's expertise with that of Professor Kober. They all met at Oxford. Ventris, just starting out as a young professional architect, did not fancy testing his wits against those of fully trained classicists. But a working relationship was formed and in the weeks and months that followed, he wrote to Professor Kober, back in Brooklyn, enthusiastically sharing his theories about the symbols.

Alice Kober sadly died in 1950, aged only forty-three. But Ventris would not give up his compulsive quest. Perhaps eccentrically, he temporarily forsook his architectural work and turned his attention to this Mycenaean mystery full-time. After having a lateral brainwave about place names to be found among the characters, Ventris found that he was pulling at a fast-opening door. Suddenly some of the symbols and characters became understandable, and at this point he found that he was dealing with an early form of Ancient Greek. By 1952, Ventris had set up a metaphorical time telescope that looked directly back to the magnificent Palace of Knossos and the daily lives that were led within it.

He was invited to give a talk on the BBC's *Third Programme*, a highly prestigious radio slot and a huge honour for a non-academic figure. This meant that his work and discoveries captured the imaginations of a large public. He was contacted by a language expert called John Chadwick, a Cambridge lecturer, who offered his assistance, and between the two of them they unfurled the entirety of the previously unknowable Linear B tablet. The achievement resonated partly because it showed that a certain mad amateur enthusiasm could succeed against the impossible. Ventris was awarded the OBE and returned to architecture. Sadly, he (like Kober) was also to die prematurely, in a car accident in 1956. But his name would live on.

And what do we know of the everyday Mycenaean world that he unveiled for us? What was the texture of life for those who had given the world such extraordinary myths? There may not have been half-man-half-bull monsters, but there was certainly a vibrant culture.

Here, after thousands of years in obscurity, were real facts from the crucible of legend. Under those ancient pale blue skies, and on that rough dusty ground, there were chariots and contests and warrior graves, and at the Palace of Knossos, there was a great king. In Knossos, there was a careful system of landholding – by women as well as men – and there were slaves immediately identified by types of garments. This was a civilisation that constantly ventured out onto the rich blue seas, not just to patrol against threats but also for the purpose of buying and selling olive oil and wine. Religion came in the form of a mysterious cult; within the Palace of Knossos was a great chamber that seemed to be there for the purpose of ceremony and worship. There were priests and priestesses and special slaves who were attached to the temple. Plus there were the Bronze Age equivalent of civil servants, keeping the palace and the city stocked with the necessities and maintaining day-to-day order.

It was out of this world that epics such as the *Iliad* had emerged, legends that encompassed the rich adventures and myths and fantastical storytelling of the time. King Minos and his monster were perhaps distorted reflections of an earlier form of animal worship (not completely unlike the later Romans and their bull-god Mithras) and, disappointingly, no evidence of a labyrinth was ever found. But what Professor Kober and Michael Ventris achieved in the twentieth century was amazing. Their codebreaking was like a form of time machine, at last enabling the world to understand more fully one of its formative civilisations. They had done so not out of obligation but out of a passion for the purest discovery. The energy with which they both taught themselves the most arcane knowledge in the end led to the unveiling of a lost time. In this sense, the cracking of Linear B created history, as well as subtly changing it.

2. THE APOCALYPSE PROPHECIES

It was an ancient language that was a gift from the gods, the means by which wisdom and science and prophecies might be conveyed. Instead of an alphabet, there were rich and varied symbols, which touched on everything from agriculture to the movements of the heavens. It was through this coded tongue that the secrets of doomsday were conveyed. This was a cipher that had encompassed an entire civilisation – and the apocalypse of that civilisation too. And ever since it was wiped out in the sixteenth century, a variety of codebreakers across the world and across the centuries have been working to decrypt the knowledge of a lost culture. The codebreaking continues today, and the object is the glyphs of the Mayan civilisation.

For centuries before the curiosity of Christopher Columbus changed the maps of the world, parts of what is now Mexico, Guatemala and El Salvador were Mesoamerica. This was the home of the Mayan civilisation. Like the Romans and the Ancient Greeks on the other side of the world, the Mayan peoples developed a rich culture, suffused with amazing architecture (ziggurat pyramids), mouth-tingling food (the rise of the cultivated chilli pepper), elaborate sculpture and a written language of elegance and complexity. As in Europe, there was a ruling monarchy and power passed down through the male line. The Mayans were also keenly aware of the importance of history, and of documenting the past. Among them were mathematicians, too, not just among the engineers who raised such astounding stone buildings deep in emerald-green forests but also among the astronomers who recorded the positions of the stars and who calculated forthcoming eclipses.

In the classic period of this civilisation, the Mayans developed thriving city states: architectural wonders of ceremonial courts, triadic pyramid temples and vast sculptures. And the literature – scholars writing on bound documents of parchment and wood – grew as well. By the time the first ominous black specks could be seen on the horizons of the oceans – those sixteenth-century ships upon which sailed Spanish explorers, who would soon be followed by rapacious conquerors – there was a body of culture that stretched back many hundreds of years, with unique features, such as a 260-day ritual calendar.

In some of the Mayan cities, the European invaders engaged in a frenzy of destruction in their quest for the vast stores of gold which they had convinced themselves were here. In other cities, more of an accommodation was reached, with uneasy truces between the indigenous people and those who sought to exploit them. But the invaders had also brought an invisible enemy with them that could not be fought off: diseases against which the inhabitants had no natural protection. When civilisations fall, they generally leave abundant traces of themselves. But as the Mayan way of life gradually dissolved over the years and centuries that followed, that record was all but destroyed: the invaders had set to work on eliminating Mayan books and Mayan language.

Catholic priests who had accompanied the conquering crews were determined to destroy the extensive texts that they found, presumably on the grounds that paganism had to be eliminated. It was a dreadful crime. Altogether too late, other Europeans realised with horror what had been done. The few texts that had somehow survived the clerical purging were preserved. And almost immediately there began a quest stretching from Middle America into the heart of Europe to rescue this lost world. But now scholars were faced with a language – the Mayan glyphs – that had become a secret code with no survivors capable of decrypting it.

They would have to try to unlock its secret meanings without cribs or help. What gave the project an extra frisson was the sense that some vital cosmic wisdom had been lost. It was (and is) believed that Mayan scholars who had observed the night skies and the universe also had unique spiritual insights concerning events to come and what might befall the world.

One of the volumes of glyphs that was saved came to be known as the Dresden Codex. Even by the time these parchments were rescued, they were of a considerable age, dating back to the tenth century or so. They were bound and contained around seventy-eight pages. For those who could unlock the language, what a potential treasure house of knowledge this was. And indeed, not all of this book was bound, some of the parchment was elaborately folded, accordion-style. Unfurled, it presented a prospect of rich and strange symbols and imagery. Fully unfolded, this document was 12 feet long. All of this had only really survived upon the whim of the Spanish conquerors; it was said that the Dresden Codex had originally been sent across the ocean as a gift to the Holy Roman Emperor Charles V. In 1739, Johann Götze, a German scholar – and custodian of the Royal Library in Dresden – bought it from an intermediary in Vienna. And it was then that the work began.

One aspect of the Mayan codes that caused a great deal of head-scratching was the ancient civilisation's system of mathematics and the way that numbers were expressed. Once this enigma was cracked, the complexities of the detailed calendars contained within the codex could also be tackled. One of the beguiling mysteries involved the 'Lords of the Night'. What part did these oft-referred-to figures play in the cycle of the hours? Gradually, over many years, the glyphs began to yield some of their mysteries. There were nine gods, each of whom ruled every ninth night, some bringing good fortune under darkness and others ill. There was a 'smoking mirror' god, a maize god and an underworld lord among them. Across the years, a noble intellectual effort to translate these otherwise inscrutable symbols ensued, with eminent scholars such as Alexander von Humboldt showing a keen and intense interest. More than this, these sorts of problems were cutting-edge exercises for more contemporary cryptology problems.

So how then was it that a few years ago newspaper headlines began trumpeting the dreadful Mayan glyph prophecy that the world would come to an end in 2012? It seems now that this was something of a codebreaking misunderstanding: the Mayan calendar, drawn up all those centuries back, came to an end in what we know as 2012, but this was prior to a cyclical reset, meaning that one long era was indeed in its dying days – but it would then be immediately followed by the birth of a new age. The story resonated perhaps for another reason: that modern (western) civilisation deserved some form of retribution for the terrible extermination of the Mayan and Aztec civilisations, and that this was the hands of their Lords of the Night reaching out across the centuries . . .

The work continues today; there is still much to be learned from the Mayan glyphs to be found in the Dresden Codex and other fragments of surviving texts. This codebreaking is endlessly rewarding, for every piece of this civilisation that can be deciphered adds immeasurably to our world's knowledge.

3. THE TREASURE OF ATLANTIS

There are some ancient codes that are like Mount Everest: there is no burning reason to conquer them other than that they are there. The mere fact of their existence teases and itches intellects through the generations, and they offer tantalising prospects of throwing wondrous light on otherwise shadowed periods of the distant past. One such enigma is the Phaistos Disc, an exquisite object of clay dating back over two thousand years which has been so cunningly wrought with mysterious symbols that it has sparked regular speculation, including hints that the coded symbols might prove the existence of a lost city of Atlantis.

A beautiful archaeological find from deep in the subterranean ruins of the Palace of Phaistos on Crete, the Phaistos Disc was discovered in 1908 by Luigi Pernier. What immediately stood out about the disc – about 6 inches in diameter – was not just its colossal antiquity but also the apparently advanced methods by which its symbols had been pressed into its clay, on both sides. These symbols, or pictograms, followed in a progression around a spiral that wound into the disc's centre. A similar design in principle was created on the reverse side too. There were 242 separate impressions of hieroglyphs around these merry spirals, and these comprised 45 distinct signs (some were repeated throughout). Once all these signs had been imprinted, the clay would have been placed into the oven to be hardened, and the creator would surely never have believed this item would continue to exist for some four thousand years.

The skill needed to create the disc would have been particular. The hieroglyphs – like the letters on a printing press – would have been solid seals pressed into the clay, so that their impression was left. But given the dainty size of the disc, these hieroglyphic seals were themselves marvels in miniature. And the geometric precision of the way that the symbols followed the spiral round into the centre could not be purely ornamental. The different hieroglyphs – those that were repeated, and those that occurred just once – surely had to have some cohesive meaning or message. But unlike other ancient puzzles such as Linear B, the disc did not lay itself open to techniques like frequency analysis.

There was a resemblance to the hieroglyphs found in the Egyptian tombs with all those cats and sideways human figures. Those on the disc differed stylistically and also because so many of the pictograms were of such particular items and archetypes. As well as cats, rams, doves and other animals being represented in quite straightforward pictorial form, there were other symbols that seemed to speak of current concerns.

Alongside pictograms for 'woman' and 'child', there was a frequently recurring image of a male head with what looked like a Mohican haircut (in fact a crested helmet). There was the head of a tattooed man; there were pictograms of manacles, and of a man in chains; there were helmets and gauntlets and slings. But pictured alongside these warlike images were other more peaceful signs: vines and lilies and bees, plus what might have been a palace floorplan, a carpenter's angle, a beehive, papyrus and a ship and water.

Taken individually, these pictograms were perfectly comprehensible, but placed into the spiral sequence of the disc, they acquired new context and significance. What was the narrative that led to the centre of the spiral? And what were the occasional short diagonal strokes that accompanied some of the impressions? Might these have been dividing points showing where words or phrases ended? Could each pictogram stand for a word, or perhaps even a phrase or sentence?

And which way round was either side of the disc supposed to be read? From the centre of the spiral whirling outwards to the edge? Or the other way round?

As the art of cryptography attained its technological phase in the mid twentieth century, it might have been expected that encryptions such as this would only register in the very dustiest and most obscure corners of academe. But this was not the case. There was something especially flavoursome and pleasing about the Phaistos Disc conundrum because of the seemingly anachronistic technical flair that had gone into producing it. Centuries before the principles of printing took off, here was an apparent early example of the art with pre-made and reuseable characters impressed on soft clay and presumably ready to be used again in all sorts of new configurations.

And was it part of the Minoan civilisation, like other curious coded finds? The symbols seemed to suggest not, though some experts argued that the pictograms were definitely more Cretan in origin. Others suggested a strong relationship with Anatolian hieroglyphs, but again there was not complete certainty. Quite recently, Dr Gareth Owens postulated that there was a religious element, having identified a reference to Astarte, a Minoan goddess of love. On the reverse side was a reference to another goddess, this one pregnant.

One of the more temptingly colourful theories prompted by the disc was that even though it was uncovered in the ruins of that particular Minoan Palace, it was already an object of antiquity when its owner had taken it there; that the disc far pre-dated the Minoan civilisation and had been kept as an impossibly old relic from another time. It was but a short step from this to the idea that the Phaistos Disc might in fact be a remnant of the long-vanished civilisation of Atlantis. Was the reason the pictograms seemed to bear little if any relation to other mysterious inscriptions, such as Linear A and B, that it had in essence come from a completely different world? The idea was more one for pleasing daydreams rather than concrete reality. How easily the thoughts drift towards that ubiquitous Mediterranean myth of a Great Flood that caused cultures to disappear. And still the probing goes on; the Disc currently has pride of place in the Heraklion Museum on Crete – a rich mystery that seems annually to yield up fresh new possibilities.

And there is some ancient precedence for special discs with religious or scientific significance. Recently, the British Museum put on show a breathtaking Bronze Age find from Germany known as the Nebra Sky Disc. Like the Phaistos Disc, there is something about the relic that can raise the hairs on the back of the neck. In this instance, it is a small circle of dark green bronze, inlaid with tiny representations of the crescent moon and the sun, with other tiny symbols. This, many thousands of years ago, was one of the very first star charts. At the time that Stonehenge was being erected, a disc such as this would have signified secret wisdom and knowledge of the year's turn, the equinoxes and how the patterns of the stars above could dictate the layouts of temples and tombs. But while the Nebra Sky Disc is now understood straightforwardly as one of the first maps of the heavens, the Phaistos Disc retains a certain elusiveness. For every theory backed up with the symbols being translated into modern binary notations, there is another giving it more mythical roots.

It remains important as a code – and an unbroken code at that – because such writings and signs serve to remind us all that however much we think we know, there are mysteries that even the most advanced computers have yet to unravel.

4. DEAD MEN AND MAGIC

On wind-bitten moors under cold north skies, the grey stones stood like men and in mists could be mistaken for them. And just as men could frequently bear facial markings as a means of tribal identification, so too the stones bore their own patterns. Even to this day, the term 'runes' sparks a sense of something Druidic, impossibly distant in time and infused with secret meanings. The landscapes of Ireland, Scotland, Norway and Sweden are still abundant in both sacred monuments and the carved coded lines inscribed upon them. But the runes, as a code system that flourished before the Latin alphabet came to northern Europe, are in fact exquisite artefacts providing illuminating insights into an otherwise time-obscured world. And runes as codes were also being used on wooden sticks, not just to convey urgent messages across long distances but also for more personal reasons too. Even the Vikings had their romantic side. 'Remember me, I remember you – love me, I love you,' ran one message on a stick excavated not long ago, near Bergen.

As an immediate proposition for deciphering, runes on first sight look impossibly forbidding (and indeed, the Victorian M. R. James's ghost story 'Casting the Runes', in which such secret symbols are used to summon up demons, might have put a few people off). The word 'rune' itself derives from various Old Saxon and Gothic terms meaning 'secret'. The vertical rune symbols, crossed with diagonals, formed full alphabets of their own – the markings denoting both the sounds of letters and words and also standing for words or terms themselves. The reason for the vertical emphasis was to do with ease of carving.

One of the best-established (and best-represented) of the rune codes was that of the Elder Futhark language. One of the more mysterious rune messages is to be found on the Kylver Stone – a graveyard marker discovered in Sweden and dating back to about AD 400. On this slab, which was used to cover a grave, there was carved an elaborate runic message. Ever since the stone's excavation in 1903, there have been efforts to find a precise decryption of what this tomb-message might have been. There were those who concluded that it contained references to horses and there was also speculation that the runes were a magical invocation to the dead man in the grave, as a means of ensuring that his spirit remained quiet.

Less romantic analysis, however, seemed to suggest that the runes referred not to the dead man at all but rather to tribes that were in existence at the time. The stone was, in a modern sense, a form of memorial. And yet that charge of ancient and awful magic remained around other runic messages. In one Viking or Old Norse poem there is a verse concerning a rune that if whispered to a dead man hanging from a tree would bring the dead man back to life so that he might walk with the poet.

Equally, there is a circle of rune-stones in Sweden with one menhir, the Björketorp Runestone, which appears to carry supernatural warnings. The main rune-stone and its carved message have been translated as meaning that anyone who damages or breaks the stone will meet a great deal of terrifying misfortune. But why would a standing stone in a field need such sinister protection? Excavations carried out failed to reveal a grave beneath, so the curse was not like those in the tombs of the Ancient Egyptian mummies. But there was a theory that the standing stone might instead have been a more generalised monument to dead people buried nearby, or even a form of shrine to the god Odin.

Yet not all rune carvings promised to bring forth unearthly supernatural vengeance. Many were more poetic than that. The most well-known rune-stone in Sweden is the Rok Stone, displaying beautiful carvings that are thought to date back to the eighth century. These runes, which cover the front and the back of the stone in neat vertical and diagonal carvings divided by long vertical lines, seem both epic and historical. They are in part a chronicle of sea-warriors, horses and battles, part showing women who offer sacrifice, and part detailing long lines of warrior sons, one of whom was fit to fight a giant. Intertwined are references to older mythology and older gods with codes carved in stone that give a tantalising glimpse into the imaginations of the far-distant dead.

And in Norse poetry, there are legends of how the runes were passed from the gods to humanity. The story of the god

Rig, who fathered three sons by three mortal women and who bestowed the secret of the runes upon the most noble of these boys, helped give that extra frisson of magical meaning to the runes themselves.

Another branch of coded runes can still be found on sites of ancient significance in Ireland, and in places in Wales, Cornwall and Scotland too. There, a form of runic language termed Ogham dated back even further. There are those who believed, in fact, that it had been in use long before the Romans made land on the British Isles. Was this, then, the lost language of the Druids? The pleasurable speculation was not matched by any available evidence, but in many ways Ogham was quite mysterious enough without needing extra layers of preternatural theory. Like its Norse cousins, the markings were broadly vertical, with short horizontal lines and diagonal slashes. The letters that they form were thought to have been named after trees: oak, hazel, alder, willow, among others.

In Munster and County Kerry are still to be found age-old standing stones, upon which Ogham runes were carved centuries ago, and some of these appeared to commemorate warriors and kings. This was also the case in Wales and Scotland. But there was also a theory that Ogham had rather more down-to-earth practical uses too, when carved upon short wooden staffs; here might have been a method of keeping tallies, or indeed passing messages between villages.

The author Robert Graves, in his epic (and singular) work *The White Goddess*, concerning the truth behind myths and legends, went a little further with his own analysis of this extraordinary code of signs. He said that it in fact dated back to the Stone Age and had emanated from a measurelessly old civilisation – long vanished – that dwelt somewhere in the Middle East on the Mediterranean coast. He believed these were 'the people of the sea' who at one point had spread out across the vastness of Europe, bringing their secret language with them. This was a language that priests and kings used to speak of matters not for other mortal ears.

In essence, this was another form of the Tower of Babel archetype. In the confusion of all the different languages of the world, there was one secret coded tongue known to special elites in lands everywhere. These runes, bestowed by the people of the sea, were taken up by Druidic priests and their ancient meanings carved into unchanging stone. But the alphabet of tree names was considered significant too, for each particular form of tree had its own spiritual meaning. As a result, there are many people today in the West Country and other regions – people who hold modern-day Druidic beliefs – who delight in Graves's theory.

There is an element of melancholy to this story too. Hermetic language that it was, the runes could not survive the spread of the new religion of Christianity, nor the widespread adoption of the Latin alphabet as the world emerged from what used to be known as the Dark Ages. By the Middle Ages, there were still some rune-carvers – the knowledge having been passed down faithfully through the generations – and in Norway and Sweden, runes were still being carved into wood in the 1400s and onto ornaments and, indeed, church bells too.

There were also some neo-pagans who darkened what had once been a tradition of knowledge. In the twentieth century, runes were notoriously adopted by the Nazi Party in Germany. The Schutzstaffel (SS) symbol was based on two sig runes, and this dreadful symbol is still to be seen in some areas of warfare today.

But of course there have been lighter uses of runes elsewhere too. The symbol for Bluetooth technology is runic. A conjoined star and angular 'B' represent the letters *hagall* and *bjarken*, which in turn are the initials for Harald Blatand, the tenth-century King of Denmark, who was known widely as 'Bluetooth'. So it is that there are some codes that reach right the way across the centuries and form part of the fabric of our modern lives.

5. THE MYTHS AND THE BATTLEFIELDS

He had written of Jason and his Argonauts, and their mythic quest for the Golden Fleece, but this Ancient Greek poet himself inhabited a land of gods and oracles. Archilochus was a warrior as well as a lyricist and had seen wonders in his lifetime. It was said that he brought the cult of Dionysus to his home island of Paros and for this the oracle of Apollo was so pleased that it demanded that the poet be honoured thereafter. Archilochus was also the chronicler of a very early form of cryptography, one of the first recorded instances of a code-carrying device being used on the battlefield. It was termed a scytale.

The world of Archilochus – who lived around 650 BC – was one of 'figs and seafaring', and warfare too. The island of Paros was relatively affluent, and those who lived there knew how to read the oceans and detect coming storms. But those storms were sometimes of the human kind, such as rivalry with the nearby people of Naxos. For Archilochus there was no contradiction between being a fearsome soldier and an emotional lyrical poet.

Only fragments remain now. Though Archilochus was hugely influential, he was also noted for his constant anger. He fought and he wrote with passion. And mentioned in these lyric poems was the encryption technique that apparently had been used by Greek generals, the scytale system. This system consisted first of two lengths of wood, both six-sided, like very large pencils. These were the scytales and both of them had precisely the same measurements and dimensions. The next element was a long thin strip of papyrus upon which were written messages. When uncoiled, the characters on the papyrus were jumbled and made no discernible sense. But when wound around the scytale in a pre-agreed way, the characters combined to make a legible message. The idea was that without scytales of exactly the same size, the enemy would not be able to read any captured messages, no matter how many pieces of wood they tried winding them around.

This method got picked up hundreds of years later by the Roman writer Plutarch. He described with admiration how the parchment was carefully wound around the scytale and the message then written upon the parchment, before being taken off, rolled up and handed to the messenger. This means of conveying intelligence across vast plains was more efficient and ingenious (and speedier) than one other idea – that of shaving the head of the messenger, writing the message on the skin, waiting for the hair to grow back to disguise it, and then sending him off on his mission. Another method that disguised rather than encrypted involved hiding messages on wooden tablets covered over with wax. Both of these ruses, termed steganography, were described by Herodotus. Their flaw was that should either the messenger or the tablets fall into the hands of the enemy, then the discovered messages would need no unravelling.

The world from which these stratagems came was one of terrific intellectual expansion. The encryption of language was in itself an extension of the fantastic literary and imaginative range of the Ancient Greeks. By modern standards the scytale is rather simple, but its very form evokes battles and sieges under Mediterranean skies, with warriors who considered it an accomplishment to compose epic poems.

1

A TO Z

In this book we will journey from the codes of ancient civilisations long past to codes looking towards the future. To start our A to Z of codes, here's an A to Z puzzle.

There are 26 letters of the alphabet. In the list there are 26 words, all starting with different letters. The challenge is to fit these words in to the grid. Words can read either across or down.

When the grid is completed, words 23 and 12 reveal vital keywords.

ARCANE BIVOUAC CODE DATES ETERNALLY FAIRIES
GRAINED HARLEQUIN INHERIT JACKDAW KERB
LASHING MIDDLE AGED NEATER OKAPI PUNDIT
QUICKTHORN REFRAIN SIGNALMAN TAHITI UNADORNED
VEGETARIANISM WEARILY XEBEC YARDARM ZEBRA

2

MAYAN MATHS

No pocket calculators, no downloaded apps, but the Mayans were formidable mathematicians. Here are some sums in which Mayan numbers are used. A shell stands for 0, a single dot stands for 1, two dots for 2, three dots for 3 and four dots for 4. A straight horizontal line stands for 5. Add one dot on top of a straight line for 6, two dots for seven, and so on. Numbers over 10 are shown by two horizontal lines, with the appropriate numbers of dots on top.

What would be the Mayan number symbols used to replace the question marks?

A + ? − =

B × ÷ ? =

C ? ÷ − =

D − + ? =

3

ALPHA TO OMEGA

Anagrams go back in time as far as the Greeks and Romans, and there were those who believed they had mystical importance.

In this puzzle the names of ancient Greek cities or city states have had the letters in their names mixed up and rearranged in alphabetical order – alpha to omega. Can you restore them to their original names?

CLUES

The list in itself is in alphabetical order, so the first three names begin towards the beginning of the alphabet.

Two names end in the same letter.

Two names begin with the same letter.

1 A E H N S T

2 C H I N O R T

3 D E H I L P

4 A A P R S T

5 B E E H S T

6 A C E H R T

4

READING THE RUNES

Read the runes by cracking the code. Each rune stands for a letter. If the first group of runes stands for A L D E R, which trees are formed by the following runes?

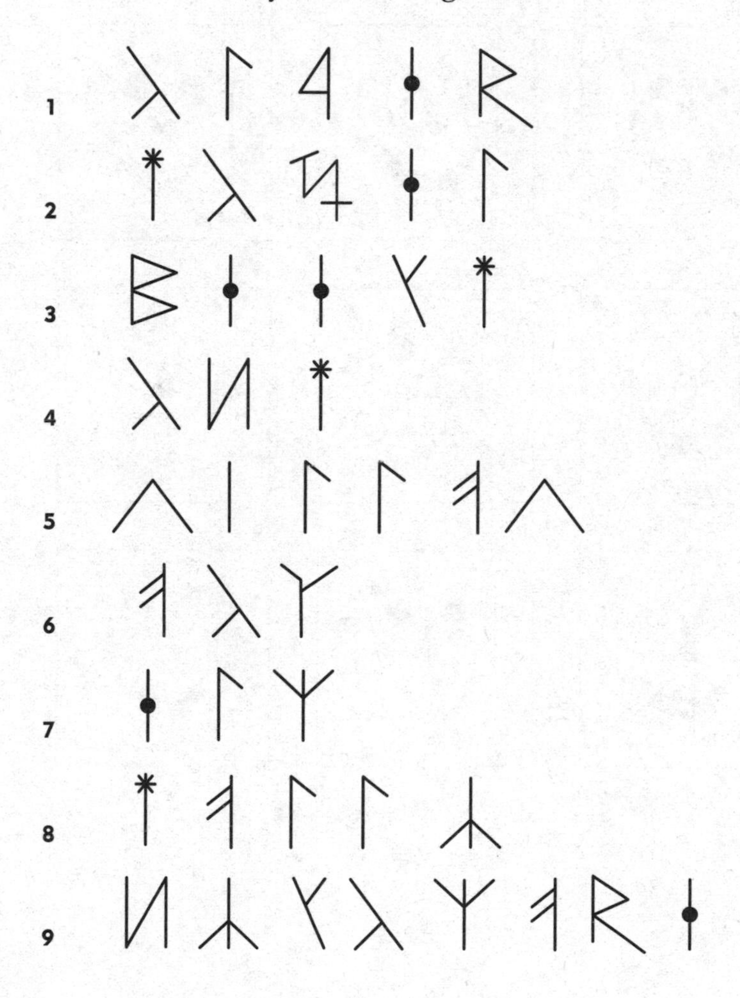

5

GLYPHS

Which of these mysterious glyphs is the odd one out?

6

SCYTALE

Wrapping a piece of parchment around a scytale was a straightforward but remarkably effective way of hiding a message.

This is a twist on the idea. You don't need a piece of wood of a precise length and thickness to sort out this message. What does it say?

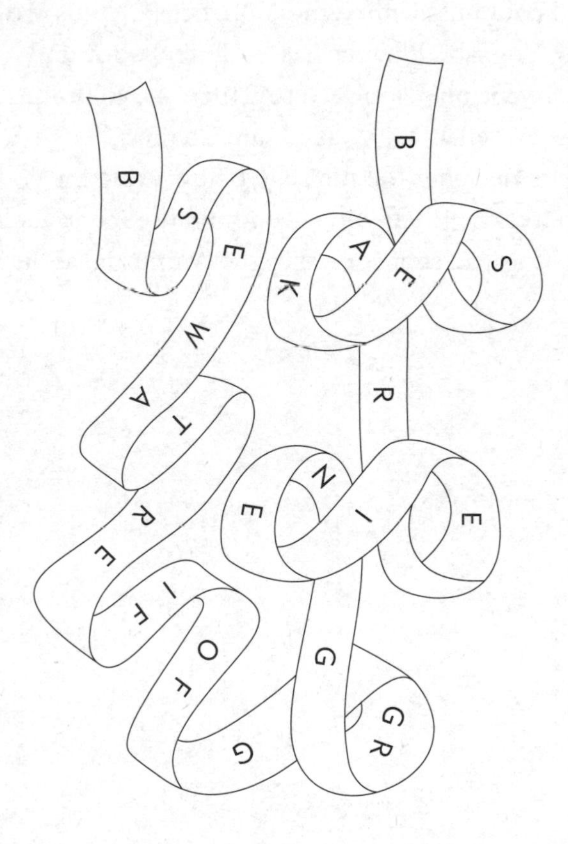

7

A GREEK GENIUS

Polybius was a Greek historian and scholar who lived in the second century BC. This cipher, which bears his name, is a simple but very effective way of writing in code using numbers to replace letters.

Here is a Polybius Square, which consists of 25 squares in rows and columns. Unlike the alphabet Polybius would have used, the English alphabet has 26 letters so X and Y share a number. To encipher, look at a row then at a column. The letter Q will be written as 42, W as 53, and so on.

Can you decipher the message below using the code of the Ancient Greeks? When you have found the message, look at it carefully. Can you spot something very unusual about it?

45.23.24.44. 42.51.24.13.31. 12.43.35.53.34. 21.35.54. 52.35.53.44. 45.35.

25.51.33.41. 35.34. 11. 32.11.55.54. 14.35.22.

	1	2	3	4	5
1	A	B	C	D	E
2	F	G	H	I	J
3	K	L	M	N	O
4	P	Q	R	S	T
5	U	V	W	X/Y	Z

CHAPTER TWO

THE CIPHER MASTERMINDS

In which we meet some of the brilliantly imaginative figures throughout the years who have changed the tides of history with their ingenious code systems.

6. ALL HAIL TO CAESAR

On epic marches through mountain valleys and across vast sun-baked plains came men in light armour carrying the emblem of the eagle. These were the armies of Caesar, cohorts of a force that held together a mighty empire. This was a world of splendour and one ruled through fear. But for Julius Caesar, brute force alone could not carry his will to the farthest reaches of his domain, from Africa to the rain-lashed British Isles. There had to be an early form of intelligence network. Thus it was that Caesar gave his name to an encryption system that created the mould for future codes. It was later to become known as the Caesar Shift and it was the means by which one of the most legendary of emperors kept his secrets safe 2,000 years ago.

The principle was simple but effective. In each encoded message, each letter would be substituted for the letter two places along in the Latin alphabet. So, A would become C, B would become D, and so on. In a continent-straddling empire with a variety of different languages and runes, and indeed illiteracy, Latin itself would have been a form of code that had to be deciphered. But the Caesar transposition gave an edge of authentic security to military and political communications. The governors of the most remote provinces with centurions under their command, attempting to quell rebellion and to keep trade flowing, needed to know what was developing in Rome. Conversely, when Caesar himself was away from the city on campaigns of conquest, there had to be a system back home that would enable his loyal lieutenants, acting as deputies, to pass word between each other without giving anything away to potential plotters. Caesar's world was filled with plotters.

'There are letters of Caesar addressed to Gauis Oppius and Cornelius Balbus, who had charge of his affairs in his absence,' wrote Aulus Gallius. 'In certain parts of the letters there are found individual characters which are not connected to form syllables but are apparently written at random; for no word can be found for these letters.'

It was the chronicler Suetonius who wrote of the system about one hundred years later with some admiration. He outlined the principles of the switched-letter system and explained the diplomatic uses of these codes. But it was Cicero – the brother of the famous orator – who admired Caesar's security in much more practical terms when he was under siege in a small fort in what is now France. His men and his fortifications were surrounded by ferocious Gallic forces called the Nervii. Despite the Romans' advanced system of creating towers and other defensive positions, the situation was growing more frayed by the day and the forests and the glades held a great deal of potential hazard. Escape or retreat would not be easy; rather, it could have resulted in bloody massacre. Yet in the midst of all this, Caesar, returning from an equally turbulent Britain, managed to get a note to Cicero via messenger, attached to a well-thrown spear. The spear and the attached message were not in fact noticed for a couple of days, but when they were, the day was saved. The essence of the message was to tell Cicero to keep good heart and remain within the fortifications, that help was on its way. This particular communication was crafted in Greek, as opposed to the substitution method, but the intention was the same. The chances of any of the Nervii being able to decipher it would have been thin.

A simple example like this does not by itself constitute an historical hinge. But the principle did, for it established another important advantage in the game of statecraft. This was a world where communication relied on messengers riding on horseback. To travel from Rome to Gaul, or indeed Alexandria, took weeks. Messages were therefore extremely vulnerable to all sorts of interception. Entrusting secrets to the few who had the code key meant building up more fundamental reservoirs of trust in those who were deputed to wield power over all the different provinces. To know of the codes was to be inducted into an inner circle of power. That trust was vital across vast territories, and across broad stretches of time.

Would Caesar have been amused to discover the different uses of his codes in the years and the centuries to follow? The answer is most likely no, if we were to cite the rash of encrypted love messages placed in the personal ads of late-Victorian newspapers. But he would have been proud that the core principles of his cipher were adopted and then adapted by increasingly sophisticated generations of cryptologists. The fact that his codes are still so recognisable across two millennia is tribute to the political mind that quickly found their most practical and effective use.

7. THE NATURAL MAGIC OF SECRET CIPHERS

A glimmering city of white and ochre, facing a bay of intense blue and in the shadow of the mighty Vesuvius, there was a point when Naples was one of the intellectual and artistic crossroads of the world. Throughout the Renaissance it was home to artists, philosophers and the earliest scientists. It was a city through which traders from distant lands brought both goods and knowledge. And in the sixteenth century, a young Neapolitan nobleman set out on a lifetime's quest to enrich the world with greater scientific understanding, and in so doing he perfected encoding techniques that are still in use today.

The young man was Giambattista della Porta, who was born into wealth and privilege in 1535. From an early age, he was encouraged to acquire knowledge. This meant, firstly, a tour of Europe, with a particular emphasis on acquiring musical skills. Upon his return to the sun-drenched Bay of Naples, aged twenty-two, he began work on the first of many books; this one being *Natural Magic*, which despite the title was less to do with magic than with 'the perfection of natural philosophy and the highest science'. He introduced it by explaining that 'if ever any man laboured earnestly to discover the secrets of Nature, it was I; for with all my mind and power, I have turned over the monuments of our ancestors and if they wrote anything that was secret and concealed, that I enrolled in my catalogue of rarities'. These 'secrets' included 'Of the Causes of Wonderful Things', 'Of Beautifying Women', 'Of Strange Glasses', 'Of Artificial Fires' and 'Of Invisible Writing'.

His first rather startling entrance into codebreaking was his adaptation of a dairy-based ruse he attributed to a second century Libyan scholar called Africanus and it involved smuggling messages to prisoners via eggs. This might sound whimsical, but this was still the time of the Inquisition, and to be suspected of heresy could be hideously dangerous. Secret messages that made it past the most intensive searches could be life-saving for many prisoners kept in dungeons. But how to avoid this stringent security? The method: write the message, using alum or plant dye, onto the shell of an egg. When the egg is dry, hard-boil it and the message on the shell will disappear. The egg could then be taken into the prison in a package of food. When it arrived, all the recipient had to do was peel the shell and they would find the message there, indelibly imprinted onto the boiled egg white.

By this stage, in the kingdom of Naples and indeed across the world, the business of ciphers was itself becoming part of a new age of scientific and mathematical enquiry. In an era riven with constant political and religious tension, with bloody conflict always close, codes were ever more vital. The problem was that the more people studied them, the easier they were becoming to crack. When della Porta was not perfecting the camera obscura – the art of using optical lenses to project what was happening outside directly to an interior screen – he was studying the cipher systems set out by Johannes Trithemius (see page 124) and some of the Arab scholars. This led to another book, completed when he was twenty-eight years old, entitled *De Furtivis Literarum Notis* ('On the Secret Symbols of Letters') – a compendium of all he had gathered from different societies about secret writing and how different systems were still evolving.

And it was in this that della Porta made his own weighty contribution to the technique of devising codes. It was not just about substituting one letter for another, or substituting a letter with a symbol, but about contriving means by which the codebreakers would be deprived of clues. He was sharp to see the importance of avoiding using the same term twice and, if possible, replacing it with a synonym (e.g., instead of repeating the word 'dog', try 'eager canine' the second time you come to use it). Equally, he was acute on how one might get a head start on deciphering a message by trying terms that might be expected to be found (for instance, in a military cipher there could be encrypted versions of words like 'soldier' or 'general' or 'commander' or 'attack'). The deciphered word – and the coded letters – could then be used to get a handle on the rest of the cipher.

And in a passage that you may need to consider when you come across a particularly head-scratching code in this book, della Porta wrote of his own experiences of deciphering that 'if the task sometimes requires unusual concentration and expenditure of time, this concentration should not go on uninterrupted', for the result of this 'excessive pain' and 'prolonged mental effort' could be 'brain-fog'. Della Porta sometimes spent so much time on codes that he 'was not aware of the approach of evening except through the shadows and the falling of the light'.

Under the languorous Naples sun, della Porta formed a prototype scientific society with a group of like-minded men – the Academia Secretorum Naturae – the aim of which, as the name suggested, was to study the 'secrets of nature'. Della Porta was assiduous in observing the life cycles of colourful flora and fauna. In fact, he was among the first in the sixteenth century to see that science had to rely upon objective experiments, where theories could be tested and measured properly. But it was still a fearful age and there were those, especially in the Church, who had a deep unease about such enquiries. They felt that God's creation was not there to be prodded and tested. Laboratories, for them, had a tang of sulphur and the occult and they believed that among those crucibles and flames were the schemes of demons. And so it was that, before long, della Porta himself attracted the hostile attention of the Inquisition.

He might have been bewildered by this frightening development; a polymath and an intellectual he may have been, but he was also a devout Catholic whose faith had never wavered. In the midst of all this, his academic society was banned and his books were withdrawn. He himself was summoned to Rome to see Pope Gregory XIII. And as author David Kahn has suggested, it is perfectly possible that della Porta was prevailed upon to dive into some cryptological work for the papal authorities.

As it was, occult suspicions were allayed and he was allowed to pursue his research. Eventually, the ban on his books was lifted as well. He joined other natural science societies and, as an adjunct to his interest in codes, he sought to invent what would later be termed a 'sympathetic telegraph'. This idea revolved around magnctism, and also around certain notions of that period to do with a mysterious 'powder of sympathy' that could supposedly heal battle wounds from great distances. The 'sympathetic telegraph' was a hypothetical communication system. The set-up: two circular boxes, each containing magnetised metal dials around which were written all the letters of the alphabet. And in the centre, a moveable arrow.

The idea was that the boxes would be magnetised by the same lodestone. Then the arrow on one of the dials would be moved to a letter – and in magnetic, magical sympathy, even at vast distances, the arrow upon the other would also move, and point to the same letter. This was a method by which messages could be conveyed almost instantly, as opposed to waiting for riders on horseback. The properties of magnetism were not yet fully understood, and it still seemed to be partly in the realm of the inexplicable. (Curiously, a parallel can be drawn with some of today's branches of quantum physics: there are twinned particles that always twitch and move in response to one another, even if millions of miles apart across space – a genuinely breathtaking idea that appears to make a nonsense of the very notion of time).

Della Porta's work on codes across the years – his version of cipher discs, with their concentric moving circles of letters and symbols, were things of beauty in their own right – was to be vastly influential. He was instrumental in taking codes far beyond the simple substitution of letters and into the realm where they needed long keys to begin unravelling them. (And while he was doing all this, he also found the time to write about twenty-five plays). Della Porta lived to the age of eighty and what he bequeathed to the world was a particular approach to secret ingenuity. In the ever more complex web of trade, science and empire that was to come, his contributions to codebreaking changed the course of communication.

8. IN THE HOUSE OF WISDOM

Born into a world of palm groves overlooking the swirling Euphrates River in what is now Iraq, there was a scholar in the ninth century who had immersed himself in other cultures and other worlds: from the texts of the Ancient Greeks to the new developments in mathematics that were coming out of India. This young man, with a resplendent name – Abu Yusuf Yaqub Ibn Ishaq al-Sabbah al-Kindi – was born in the city of Kufa, where his father was the governor. In his youth, he was sent off to study in Baghdad. As well as embracing music and philosophy and the rich resonance of poetry, Al-Kindi was one of the very first people to write a detailed treatise on the science of cryptography. His influence would subsequently be felt across the centuries, and yet, paradoxically, his name would be forgotten for an equally long time. For even though he moved through the scented palaces of mighty caliphs, his life was to end in sad obscurity.

As a young man, Al-Kindi was inducted into the House of Wisdom in Baghdad. This was an extraordinary institute that encompassed the world's learning, and in which scholars were expected to expand their intellects into the full range of human knowledge. It was one of the crowning achievements at the start of what is termed the Islamic Golden Age; Baghdad in the ninth century was a world city, gathering travellers and merchants from distant lands. And the House of Wisdom was said to be one of the greatest repositories of knowledge from all these lands, from ancient philosophy to the latest observations in astronomy, to the art of music. There were also scholars who practised medicine and those who were expert engineers. And all this took place under the eyes of caliphs who saw learning as one of the greatest treasures – as these were rulers who themselves had scholarly inclinations. For Al-Kindi, entrance into the House of Wisdom was also a gateway to meetings with more seasoned poets and astronomers. It also gave him his invitation to work for the caliph Al-Ma'mun.

Among Al-Kindi's talents was that of calligraphy. The hypnotic swirls and curves of Arabic writing required a particular elegance of hand. But he was also sought after as a philosopher; here was a man who could mesh the ancient world with the thriving new Islamic age. It was a life of worship, study and beauty. In this city, encompassed within a vast white circular wall, were domes and minarets, sand-coloured houses, tall palms, bustling markets and cooling waterways. Al-Kindi spent a great deal of his time translating Greek scientific manuscripts into Arabic for his caliph. He also took a keen interest in the system of numbers used in India, turning these into Arabic numerals, which are those – 1 to 10 – that we use today.

In other words, Al-Kindi, and his city, Baghdad, were at the crossroads of the world. All knowledge flowed there, and it was absorbed, adapted and used, and in turn was sent back out into the world. It was perhaps only natural that such an intellect – working closely with language and numbers, in proximity to the caliph – should also become one of the first experts on cryptography.

He was the author of a study entitled *On Extracting Obscured Correspondence*. There had been an earlier Arab scholar called Al-Khalil who had begun exploring the labyrinth of codes. But Al-Kindi plunged deeper, and with delight. He was among the first to see that frequently used letters would help to start unravelling codes – as in this sample plaintext phrase 'pleasing leaps in Paris' with its abundance of 'p's, 'i's, 'a's and 's's. The encoded letter 'p' – which might be turned into an 'h' – would still appear three times in the phrase. Al-Kindi reasoned that one could draw up a table of the encoded letters that appeared most frequently and begin by a process of elimination to try to piece together the jumbled words. He moved nimbly between Arabic notation and the Roman and Greek alphabets.

Arabic is incredibly voluminous as a language, with a vast vocabulary that is detailed and precise and rich. In fact, they have at least fourteen different words for 'love'. It was likely because of this that Al-Kindi found his skill with cryptography sharpened and honed more than others'. As Arab influence spread beyond the Middle East in the medieval period, so too did the lustrous labyrinthine complexities of encoded communications. Arabic was the language of maths and science, and the spread of Arabian rule over Spain at the start of the Middle Ages brought with it whole new reservoirs of knowledge.

This was in part down to pioneers like Al-Kindi, poring over innumerable parchment books in the cool surrounds of the House of Wisdom. He himself would later become a tutor to the caliph's son, but then royal death and dynastic changes altered the political temperature in ninth-century Baghdad, and Al-Kindi found himself being muscled out of the House of Wisdom by scholarly rivals. In spite of the vast amount of literature that he had produced – as well as codes, he expounded upon subjects as diverse as meteorology, music and metaphysics – his works were abandoned and he was gradually forgotten, doomed to sink into the shadows of the city.

And he remained forgotten for a long time, yet like a fire that refuses to die out, flickers could still be seen across Europe. Some of his works were translated into Latin, and he began to become known as one of the founding giants of Arab thought and culture. Copies of some of his many manuscripts materialised in what is now Istanbul, which sparked an intrigued reappraisal.

More particularly, Al-Kindi's voracious intellectual appetite, which led him to explore the linguistic and mathematical possibilities of codebreaking, were extremely influential. His ideas resonated down the centuries and across the great trading routes of the world. There will undoubtedly be many who enjoy cryptography today who would give anything to be able to jump into a time machine and visit the House of Wisdom.

9. THE MIND-TWISTING MAGIC SQUARE

The stories of kings are one thing, but sometimes the lives of those who served them are more gripping yet. This is especially true of trusted aides who were not born into nobility, and who seemingly rose from nowhere. One famous example is Thomas Cromwell who, in the early 1530s, became the apparently indispensable right-hand man to Henry VIII. Indispensable, that is, until he wasn't. Cromwell's humble beginnings in Putney, and his formative years moving around the Continent, acquiring languages, knowledge and wisdom until finally he found a role at the heart of Henry's royal court, engaged in the highest affairs of state, seem almost a parable of social ascension.

In France, just a few years later into that century, there was a man who didn't quite achieve the dizzy heights of power that Cromwell held and yet, in his own way, through his work with dukes and with France's King Henry III, left a mark on the world as distinctive as Cromwell. Blaise de Vigenère was a skilled diplomat, and an author and historian; he was also mesmerised by cryptography. And as such, he came to invent a brilliantly simple – yet extraordinarily difficult – cipher system that was used for centuries thereafter. What is also interesting is that this system and its creator were at the centre of a particularly fraught historical moment in Europe, with religious conflict tipping over into bloodshed right the way through the Holy Roman Empire.

Blaise de Vigenère was born in pretty much the geographical centre of France, the village of Saint-Pourçain-sur-Sioule, in 1523, just as the shocks and aftershocks of the Protestant revolution were being felt in nations across the Continent. The family was moderately well-to-do, although very far from being socially exalted. But de Vigenère's mother saw his intellectual potential, and so he was thoroughly schooled in Greek, Hebrew and Italian, and his education took him to Paris. One of his tutors was the poet Jean Daurat, whose work was infused with ideas about the reformation of the language.

Such an education led logically to the diplomatic service,

where de Vigenère began his long career in 1549, at the age of twenty-six. It was not long before his duties became more heavyweight. In one instance, he and the French envoy attended the reconvened Diet of Worms (most emphatically not a diet of worms, but in fact a form of political and religious assembly in the city of Worms, which was called at times of transnational tension). He became secretary to the Duke of Nevers and for a time was seconded to the service of Henry III. All of these positions would have required discretion and secrecy in communications, and it was while de Vigenère was seconded for a couple of years to Rome that he began giving serious thought to the science of ciphers.

He met with cryptographers and studied existing systems, such as the code discs of Leon Battista Alberti (see page 49) and the refined ciphers of Trithemius, so eagerly taken up by Dr Dee (see page 123). He also became absorbed by the work of one Giovanni Battista Bellaso, who had a system of switching between cipher alphabets for every letter. The very basic principle was like the transposition codes of Caesar but it was given additional knots of difficulty by multiplying the number of ciphered possibilities for each letter. This was (and is) termed polyalphabetic.

But how could this work as a fast, practical, portable – and yet super-discreet – cipher system? Blaise de Vigenère's idea was one of evolution. Out of previous ciphers (mainly that of Bellaso), he envisaged a large square. This was the basis of an 'autokey cipher': the square was a grid of 26 × 26, and was filled with the letters of the alphabet horizontally and vertically, so that in the first row and column, A to Z would run in both directions, and in the next, it would run with B as the starting letter of the sequence, so that by the end of the line, the penultimate letter was Z and the final letter was A. With each row, the sequence of the alphabet would be shifted along by one letter. One could then read off the letters to be encrypted rather like a map reference, with alphabets running along the top and down the side.

What was then needed was a keyword: 'Pope', for example, and a message, such as 'Kill them now'. The keyword and the message would be run together, so it would read: 'Popekillthemnow'. With keywords agreed between sender and recipient, codes would be formed using the grid and indeed unravelled using the same grid. The method was quite straightforward, involving reading horizontally along the row and then vertically down the column to establish the meeting point of the two lines. This is where the crucial letter would be found. Since the keyword only had to be short and pithy, its security was reasonably tight. Nor did any keyword ever have to be reused. With each fresh encryption could come a completely new keyword.

In 1586, the by now middle-aged de Vigenère published a treatise on this cipher, which he presented to the French King Henry III. And even though it was more properly the inspiration of Bellaso, the term 'Vigenère Cipher' stuck. The code also fast gained a reputation for being wholly, genuinely uncrackable, for unless one was able to guess at the keyword, the multiple substitutions and resubstitutions of every single letter would not yield to any logical method that anyone could think of.

Europe was on the thunderous brink of a dreadful period of darkness: the Thirty Years War of the 1600s caused uncountable deaths across the Continent. And between all those warring courts and factions were encryption systems that had sprung from the professional fascination of a French diplomat who had gone on to publish twenty other books on a whole range of subjects. More than this, the code was to cross oceans. It was in use throughout the bloody American Civil War, where Confederate and Unionist soldiers and officers had – instead of paper squares – brass cipher discs which carried out the same kind of shift function. It was after that war ended that mathematicians started to take an interest in deconstructing the square and finding means to attack it through pure logic.

In 1863, Friedrich Wilhelm Kasiski – a German codebreaker and archaeologist – felt ready to launch his own assault on what had, for the best part of three centuries, remained a locked enigma. His method began with a mathematical deduction of the length of the keyword. Once that had been established, then other elements of the code could be subjected to frequency analysis, searching for commonly occurring letters. There was also the chance that repeated words within the message might have repeated encrypted letters. Some sixty years later, the American code genius William Friedman (see page 284) applied himself to the Vigenère challenge and went further in formulating mathematical equations using, in part, theorems to do with coincidence and probability, concerning the frequency with which random encrypted letters might recur.

All of which is a tremendous tribute to de Vigenère (and those who inspired him); he popularised a code that took the best part of three centuries to crack, and the best part of four centuries to properly analyse its tremendous mathematical depth. And in the meantime, uncountable numbers of military messages and diplomatic communications remained secure thanks to a young Frenchman's simple zeal and passion for the subject of cryptography.

10. THE WHEEL OF DESTINY

In the era of Leonardo da Vinci, genius was not compartmentalised. If one was a great artist, then there was always a possibility that one was a great mathematician as well. In the fifteenth century, over-achievement could look commonplace. And the greatest example was Leonardo himself: what could be more natural than for the painter who could summon ethereal beauty with his brushes to also invent – and design – an early form of helicopter?

Similarly, a young man born in Genoa in 1404 would not only go on to flourish as an architect and artist and athlete, with an influence as wide as that of Leonardo, he would also become a key figure in the development and history of cryptography. Leon Battista Alberti would have understood that all of these pursuits had deep philosophical connections. Everything did back then. The seeds of his passion for cryptography were sown when he was very young. Even though he was illegitimate – his father was a wealthy businessman, his mother an unnamed figure from Bologna – he was cherished and fully recognised. The family (he had acquired a stepmother) moved to the elegant waterways of Venice and Leon was sent to a prestigious boarding school in Padua that placed particular emphasis on literary dexterity.

The boy became such a whizz at Latin that he perpetrated a rather brilliant hoax: the 'discovery' of a comedy by the Roman statesman Marcus Lepidus called *Philodoxeos Fabula* – or 'Lover of Glory'. Naturally, the long-dead Lepidus had nothing to do with it; this sophisticated work was entirely produced by a precocious schoolboy. Yet even as his intellect blossomed, calamity struck. Leon's father died, and the illegitimacy of both he and his brother now made them vulnerable to other hawkish family members who wanted them excluded from the clan's fortunes. Leon Alberti's life, which had been one of unthinking material security, was suddenly precarious: how was he to forge ahead amid hostile relatives?

But this was a time in which supremely talented young men could find advancement by means other than simply family money and connections. First, Leon Alberti studied law, though he found it insufferable. He then began to specialise in mathematics and literature, while finding work as a secretary to various noblemen. Such was Leon's energy and will to succeed that he even taught himself music, and when not cloistered away in his studies, he enjoyed mountaineering with friends. He was such an accomplished athlete that it was claimed he could leap right over the heads of said friends.

Perhaps inevitably, he was drawn into the orbit of the Church. Leon went to work for Bishop Molin and ended up taking holy orders himself. As well as being a passionate expression of his faith – he wrote and painted lives of the martyrs – this also gave Alberti a secure income, which in turn allowed him to continue pursuing his other highbrow interests. Great works of art and architecture were to follow. He spent time in Florence, where he and Leonardo da Vinci had many opportunities to trade their extraordinary, dazzling and innovative ideas. Eventually, a contemporary of his became Pope Nicholas V, and Leon was appointed to oversee and direct an amazing new period of architectural brilliance with churches and chapels.

It was his fascination with the amazing new technology of the moveable-type press that inspired Alberti's immortal foray into the realm of cryptography. He came up with an idea that would later be adopted and adapted by Trithemius and Dr Dee: that of the cipher wheel. What made Alberti's conception so eerily ahead of its time was that this brilliant wheel shared a key characteristic with the Enigma machine that would follow some four centuries later. Here was a wheel within a wheel: a disc that stayed still and another that moved, both divided up into the letters of the alphabet. The pre-agreed key, for instance K, and the first letter of the message, lined up together so that all the other letters could then be read off above their unencrypted equivalents, would dictate the way the coded message started. The ingenious part was that the pre-arranged key would change with the first letter of each new sentence, so in essence the entire code changed thereafter. And it also meant that it was no use looking out for repetitions of common letters such as E, which might have been turned into a Q, for instance; in previous codes, the E would always be encrypted as a Q for the rest of that message. Leon changed all that. The change of the keys, and the turn of the cipher discs, ensured that E was regularly encrypted as something else. On top of all this, the cipher discs looked rather beautiful and they were simpler to use than the Vigenère Square.

But Leon might have been surprised that he was remembered for a cipher, no matter that it was to all intents and purposes unbreakable, and therefore measurelessly valuable to all sorts of nobles and generals in the years that were to follow. It was just one of a myriad of ideas and inventions that sparked from his imagination. There was one incredible wheeze – part mathematical brainstorm, part prophecy of robotics – which allowed a mechanical arm, a thread and a weight to take three-dimensional measurements of statues. There were scholarly essays on matters such as perspective and geometry in painting. There was even an invention for measuring the exact depth of vast ocean waters. For a man who left his imprint on art and architecture and maths, it was only natural that he should also leave the world an immortal method of creating unbreakable ciphers.

1

GAME PLAN

In a cipher slide, the individual letters in the plaintext words are all moved forward in the alphabet. So, let's take the word SLIDE. If every letter moves three spaces forward in the alphabet, S becomes V, L becomes O, I becomes L, D becomes G and E becomes H. The word is written as VOLGH.

There are some coded word games still played today that Roman chidren would have been familiar with.

Here are the words that have been enciphered. There are four games and two pieces of playground equipment. How many alphabetical places have the letters moved? That's for you to solve!

CLUE: The slide move is less than ten places.

GAMES

1 Z G M

2 N U V Y I U Z I N

3 R K G V L X U M

4 N O J K - G T J - Y K K Q

EQUIPMENT

5 Q O Z K Y

6 Y C O T M

2

MUDDLED MATHEMATICIANS

The names Pythagoras, Euclid and Archimedes are well known as giants in the field of mathematics. In the code puzzle below, the mathematicians have got a bit confused with their calculations. Or have they? There is a code to make all these sums constant and correct. What should replace the question mark in sum 8?

CLUE: There is a code to make all these sums constant and correct. The key? Think of numbers written as words – and also about Roman numerals. What should replace . . . ?

1 5 + 7 = 9

2 8 + 9 = 2

3 10 + 7 = 5

4 6 + 5 = 13

5 6 × 10 = 0

6 7 × 6 = 45

7 5 × 7 = 20

8 9 × 5 = ?

3

CAESAR SALAD

It would have been a wonderful, lasting tribute to Julius Caesar if he had invented the world famous Caesar salad. Unfortunately, it was nothing to do with the emperor but instead the work of chef and restaurant owner Caesar Cardini in the last century.

The letters in the name CAESAR – that's A, C, E, R and S – have been removed from these foodie items. Other letters remain in place. All the items could appear on a menu today. Can you work out the names?

1 _ _ _ _ _ _ O L _

2 _ _ U _ _ G _

3 _ _ _ _ L O P _

4 _ T _ _ K   T _ _ T _ _ _

5 W _ T _ _ _ _ _ _ _

6 P _ _ _ H   M _ L B _

7 _ H _ _ _ _ _ _ K _

8 I _ _   _ _ _ _ M

4

INVISIBLE WRITING

Giambattista della Porta devoted much of his life to the study of codes. His curiosity and thirst for knowledge led him to write about many different ideas, including the intriguing concept of Invisible Writing.

What is the message from these three boxes of letters?

2

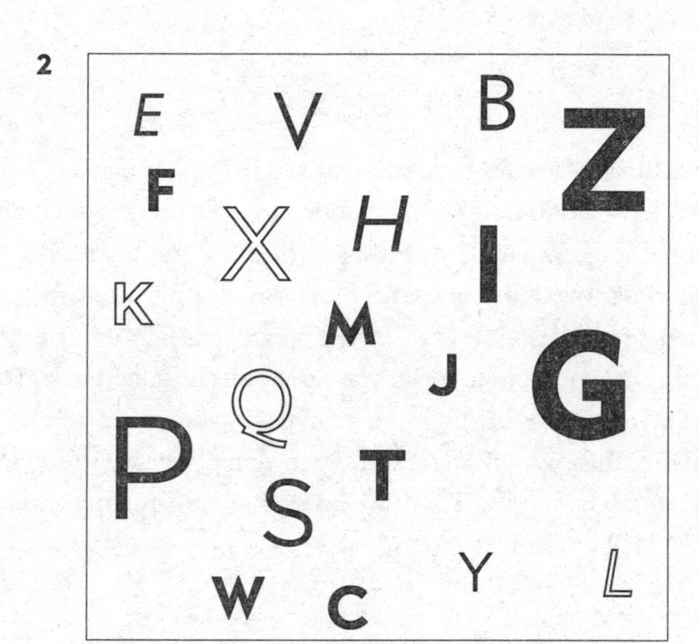

3

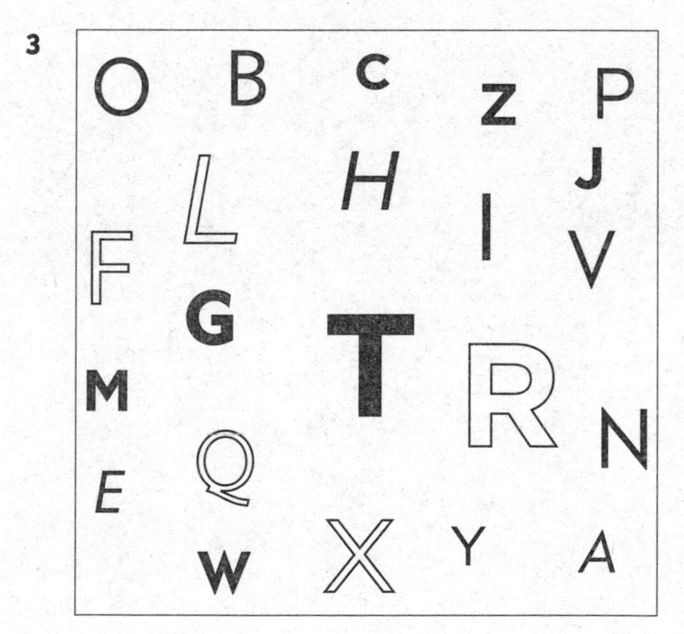

5

PIGPEN

The history of the Freemasons evolved from organised groups of working stonemasons to a modern system of regional and national lodges. Much secrecy surrounded the brotherhood and codes proved to be a useful means of communication. The pigpen cipher, also known as the Freemasons' cipher, is made up of a chart or table of letters and dots, all contained in square or triangular 'pens'. Each letter of the alphabet is represented.

Using the pigpen chart, can you decipher the following message which begins, 'There are no strangers in Freemasonry...'

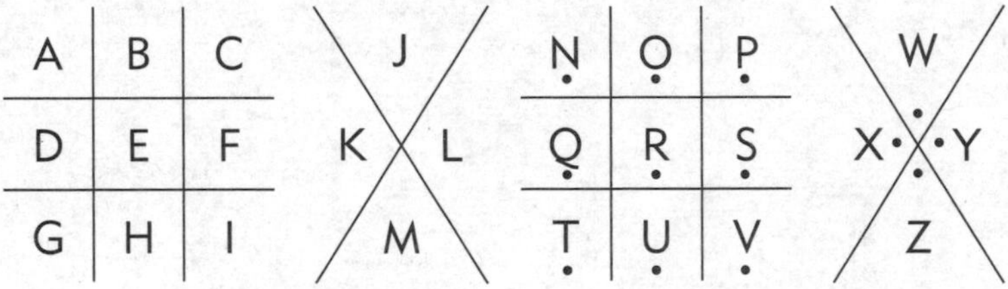

A B C
D E F
G H I

J
K L
M

N O P
Q R S
T U V

W
X Y
Z

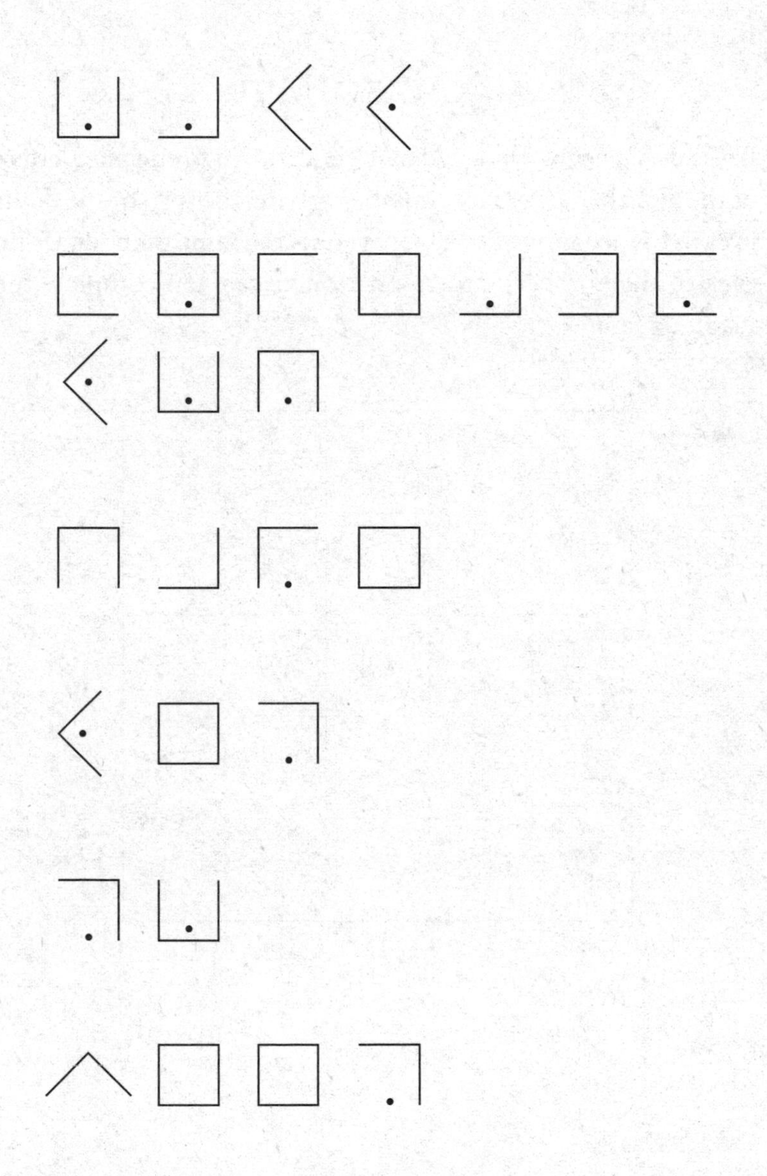

6

MAGIC SQUARE

Blaise de Vigenère produced his magic square to encipher letters of the alphabet. In a magic number square, the sum of the digits in each row, column and diagonal equal the same number. Using each of the numbers 1 to 25 only once, can you complete the magic square?

			3	
8	14		21	
		13	19	25
24				18
17		4		11

7

WHEELS

The cipher wheels of Leon Battista Alberti featured a moving circular wheel fixed inside a larger wheel. Both wheels had the alphabet printed around the outside. A key letter could be selected and the inner wheel moved so that the letter A was directly below it. All the plaintext letters on the inner wheel could be replaced by the corresponding outer wheel letters. The beauty of the system was that the key letter could be replaced as often as the encoder wanted.

Five different animals have had their names encoded. A new key letter has been used for each word. Can you work out the names? The five key letters will spell out the name of another animal. Happy hunting!

1 E B H G

2 X W V F

3 H K G X

4 D I F V E

5 Y E H I V

8

WISE WORDS

Studying in the House of Wisdom, Al-Kindi noticed the link between repetition of words and letters and realised that this could provide a way in to breaking a code. Look at the paragraph below.

Students seeking to solve ciphers require unusual enthusiasm.

Using guile, guesswork and questioning, seemingly meaningless messages and signs, genuinely seem to make sense.

What does 6.1.3.4.5.2. spell out?

9
THE VIGENÈRE SQUARE

As well as being a majestically clever means of turning language inside out, in a seemingly infinite variety of ways, the Vigenère Cipher – or Autokey Cipher – is also a thing of beauty.

The giant square – the *tabula recta*, as found overleaf – takes the letters of the alphabet and puts them through sliding permutations. So when faced with this, how do you begin to retrieve messages from it?

Here is a very slightly streamlined version of a cipher system that was in use for centuries. If you are forearmed with the code-key, the process should pick up speed as you go along.

In the first code given overleaf, the keyword is 'Shakespeare'. Here is the method: in the manner of the Caesar grid (see p.66), write the letters of the keyword above the encrypted letters. So, for instance, for that first group of four encoded letters –EPSC–, you will write the letters 's', 'h', 'a' and 'k' above them. The next group: 'e', 's', 'p', 'e' and so on. Once you get to the final 'e' of 'Shakespeare', start again with the 's'.

Now you are ready. Track down to the 'S' row – and then track along that row until you hit the letter 'E'. From here, track up to note the letter at the top of the column. You should find that it is 'M'. This is the first unravelled clear letter.

Do the same with the second letter of the keyword: 'h'. Track down the first left-hand column to row H, then move your finger along it until you hit 'P'. Glance up to the letter at the top of that column and you will see 'I'. This is your second decoded letter. From here on in, the unravelling should get easier.

And to devise your own codes, reverse the process. Write out the message, and then the keyword below. To find the first letter of the code: if the word, as here, is 'missile', then track down the first column to reach the 'M'. Run your finger along the row until you hit the letter in the S column, since 'S' is the first letter of the keyword ('Shakespeare'). The letter you are pointing at will be 'E'. Do the same with the second message letter, in this case 'i'; move horizontally along the I row until you reach the letter in the H column (H being the second letter in 'Shakespeare'). That letter will be 'P'…

And as you do so, you will have the satisfaction of knowing that you are using a cipher system that stretches back to the Renaissance. There is even greater satisfaction in constructing ciphers with it.

CODE TEASERS:

1 EPSC MDTF AJIZ PDNI FJRD VVEV OB
KEYWORD: SHAKESPEARE

2 NPYX FOAN BGUT HZLEAHB
KEYWORD: BLUE DANUBE

3 LATQ TNFJ JMRJ GNQP XEXW ON
KEYWORD: ST PETERSBURG

4 MIXM ITGY NSPP JJWIH
KEYWORD: LULLABY

5 LRDV WUOH VJZB CETH FDRG LA
KEYWORD: RADIOACTIVITY

	A	B	C	D	E	F	G	H	I	J	K	L	M	N	O	P	Q	R	S	T	U	V	W	X	Y	Z
A	A	B	C	D	E	F	G	H	I	J	K	L	M	N	O	P	Q	R	S	T	U	V	W	X	Y	Z
B	B	C	D	E	F	G	H	I	J	K	L	M	N	O	P	Q	R	S	T	U	V	W	X	Y	Z	A
C	C	D	E	F	G	H	I	J	K	L	M	N	O	P	Q	R	S	T	U	V	W	X	Y	Z	A	B
D	D	E	F	G	H	I	J	K	L	M	N	O	P	Q	R	S	T	U	V	W	X	Y	Z	A	B	C
E	E	F	G	H	I	J	K	L	M	N	O	P	Q	R	S	T	U	V	W	X	Y	Z	A	B	C	D
F	F	G	H	I	J	K	L	M	N	O	P	Q	R	S	T	U	V	W	X	Y	Z	A	B	C	D	E
G	G	H	I	J	K	L	M	N	O	P	Q	R	S	T	U	V	W	X	Y	Z	A	B	C	D	E	F
H	H	I	J	K	L	M	N	O	P	Q	R	S	T	U	V	W	X	Y	Z	A	B	C	D	E	F	G
I	I	J	K	L	M	N	O	P	Q	R	S	T	U	V	W	X	Y	Z	A	B	C	D	E	F	G	H
J	J	K	L	M	N	O	P	Q	R	S	T	U	V	W	X	Y	Z	A	B	C	D	E	F	G	H	I
K	K	L	M	N	O	P	Q	R	S	T	U	V	W	X	Y	Z	A	B	C	D	E	F	G	H	I	J
L	L	M	N	O	P	Q	R	S	T	U	V	W	X	Y	Z	A	B	C	D	E	F	G	H	I	J	K
M	M	N	O	P	Q	R	S	T	U	V	W	X	Y	Z	A	B	C	D	E	F	G	H	I	J	K	L
N	N	O	P	Q	R	S	T	U	V	W	X	Y	Z	A	B	C	D	E	F	G	H	I	J	K	L	M
O	O	P	Q	R	S	T	U	V	W	X	Y	Z	A	B	C	D	E	F	G	H	I	J	K	L	M	N
P	P	Q	R	S	T	U	V	W	X	Y	Z	A	B	C	D	E	F	G	H	I	J	K	L	M	N	O
Q	Q	R	S	T	U	V	W	X	Y	Z	A	B	C	D	E	F	G	H	I	J	K	L	M	N	O	P
R	R	S	T	U	V	W	X	Y	Z	A	B	C	D	E	F	G	H	I	J	K	L	M	N	O	P	Q
S	S	T	U	V	W	X	Y	Z	A	B	C	D	E	F	G	H	I	J	K	L	M	N	O	P	Q	R
T	T	U	V	W	X	Y	Z	A	B	C	D	E	F	G	H	I	J	K	L	M	N	O	P	Q	R	S
U	U	V	W	X	Y	Z	A	B	C	D	E	F	G	H	I	J	K	L	M	N	O	P	Q	R	S	T
V	V	W	X	Y	Z	A	B	C	D	E	F	G	H	I	J	K	L	M	N	O	P	Q	R	S	T	U
W	W	X	Y	Z	A	B	C	D	E	F	G	H	I	J	K	L	M	N	O	P	Q	R	S	T	U	V
X	X	Y	Z	A	B	C	D	E	F	G	H	I	J	K	L	M	N	O	P	Q	R	S	T	U	V	W
Y	Y	Z	A	B	C	D	E	F	G	H	I	J	K	L	M	N	O	P	Q	R	S	T	U	V	W	X
Z	Z	A	B	C	D	E	F	G	H	I	J	K	L	M	N	O	P	Q	R	S	T	U	V	W	X	Y

10

THE CAESAR SHIFT

This system, as used throughout the era of the Roman Empire, was an early form of transposition cipher. In realms not versed in Latin, it must have seemed dizzying. But the principle is very simple indeed. The top row of letters in the grid show the alphabet. The bottom row shows that same alphabet, but with each letter moved two positions along to the right, so that 'A' falls under 'C' and 'B' under 'D', etc.

The messages below have been encoded using this grid. It will not take you long to unravel them. If you were a commander in charge of a garrison under attack, this speedy method might have been a relief.

And when you want to construct your own ciphers, the grid simply works in reverse.

A	B	C	D	E	F	G	H	I	J	K	L	M	N	O	P	Q	R	S	T	U	V	W	X	Y	Z
Y	Z	A	B	C	D	E	F	G	H	I	J	K	L	M	N	O	P	Q	R	S	T	U	V	W	X

1 RFC PCZCJ EYSJQ FYTC ZCQGCECB SQ

2 AYL ZPSRSQ ZC RPSQRCB?

3 RFC FMPBCQ YPC YR RFC JMLBGLGSK EYRCQ

4 AJCMNYRPY FYQ QCYJCB Y QCAPCR YJJGYLAC

5 MSP DJCCR FYQ QYGJCB RM RFC CBEC MD RFC UMPJB

CHAPTER THREE

THE BIBLE ENIGMAS

In which we explore some of the more encrypted corners of the Old Testament, and the terrifying visions therein that were used as codes.

11. THE NUMBER OF THE BEAST

Throughout the centuries, religion and secret codes were frequently intertwined. Sometimes it was because of cruel, violent persecution. Sometimes it was because senior priests wanted to jealously guard sacred knowledge, protecting it from dangerous unbelievers. But in the Bible, the most intense and colourful of codes are to be found – for all to read – in the Book of Revelation. The account of the coming apocalypse remains incredibly pervasive today: the Four Horsemen are still recognisable emblems of terror, and the mark of the beast, the number 666, remains both factual and fictional speculative territory right the way around the world. That this final book of the Bible is reckoned to have been written by one John of Patmos around AD 95, making it almost 2,000 years old, qualifies it as one of the most durable and nightmarish of all cryptographic propositions. For there is no question that the feverish and terrifying images throughout are intended as such.

What was it all supposed to signify? Who was the beast, and why did the number 666 carry such particular significance? Who or what was the Great Whore of Babylon? And what was the reason behind so many recurring numbers – such as the seven seals, the seven horns and the seven eyes? This was a secret code that was intended to be both universal and yet also very local to the territories around the eastern Mediterranean. And like all the best secret ciphers, there is speculation that its true intention was to spark the overthrow of imperial rule.

Reading the Book of Revelation has been likened to taking drugs and experiencing some particularly nauseous hallucinations. It begins calmly enough with John having prophetic visions, which he is commanded to convey to the 'seven churches of Asia' (the fledgling Christian churches in various realms around the Middle East and the Mediterranean). There is an image of seven stars and of seven lamps too. But then we are before the Throne of God, where a scroll with seven seals is to be opened by a lamb with seven horns and seven eyes. The opening of the first four seals brings forth horses: a white horse bearing a rider with a bow; a red horse bearing a rider with a sword; a black horse, the rider of which is holding a pair of scales; and then, rather more darkly, a pale horse, whose rider is Death. It is all downhill from this point onwards.

The opening of the seventh seal brings forth seven trumpets, the sounding of each of which summons fresh horrors. There is fiery hail and rain of blood that suffuses and burns the grass and trees, and a mountain falling from out of the sky and into the ocean. A sinister star called Wormwood shoots from out of the dark heavens and crashes into the earth, poisoning the crops, and the sun and the moon are plunged into darkness. Then giant locusts 'with the power of scorpions' appear from out of the earth. These monsters have human faces and hair, lions' teeth and vast wings. There are a few among mankind who will find salvation, but multitudes who will not.

The imagery multiplies dizzyingly and amid all the thunderstorms and the cracking open of the sky a pregnant woman in white, with 'a crown of twelve stars', is to give birth to a boy. A terrible dragon, or devil, with ten horns and seven heads, is awaiting this birth greedily, for he intends to eat the child. But the woman in white escapes the dragon, and in turn, this brings war to Heaven, with the dragon being fought by Michael. Elsewhere, a beast materialises from the earth with two horns and the voice of a dragon. This is the creature that is known by his number – 666.

After all these nightmarish doomsday images, the powers of darkness are eventually destroyed and the ecstatic harmony of New Jerusalem is established. The promise of salvation is held out. But the question that has been debated by theologians and historians for the best part of 2,000 years is: what exactly did John of Patmos mean by it? If it was not a literal prophecy – and it was too labyrinthine to really be understood as one – then how should one decipher the author's true meaning?

The haunting images, it has been observed, were also sometimes found in older Jewish literature. But the demonic world that is being depicted may well be, as some theorise, a coded version of the reality that John of Patmos saw around him. The pagan excesses typified by beasts, blood and whores were his enciphered means of describing the oppression of Roman imperial rule. Some of it was to do with martyrdom, acknowledging the early Christians put to a horrible death for their faith. But surprisingly, expert economist Ian Smith, in a recent interpretation of the Book of Revelation, reads the code as being (partly) about early Christians in cities and towns losing work and having their trade shunned and thence being tempted into worshipping pagan idols, thus condemning their souls.

The trouble was that in the Roman Empire at that time, the emperor himself (Domitian) was set up as a sort of god. Temples in the city bore his image. And as a pagan idolatrous religion, it was also very clubby. 'Participation in the cult of the emperor,' writes Smith, 'with its festivals, temples, altars, images, games, processions and public meals was attractive, since social and political advantages and status depended on ties to the cult.' But, according to John, Christians who gave in to this idolatory were backsliders rather than craftsmen who needed to socially network. Those who held out would taste the bliss of paradise. Thus the beast that rose from the sea was code for Roman military might, the beast that arose from the earth was the 'cult of the emperor', and that oft-mentioned Whore of Babylon was Rome herself.

In this sense, the apocalypse was partly a cipher, partly a prophecy (for woe betide those who joined in with pagan rites) and partly a seethingly furious diatribe against the iniquities of the world in the first century AD. Yet here was a code that changed the religious imaginations of generations to follow and influenced outbreaks of unrest and fear 1,000 years later as the end of the first millennium drew close and believers imagined that the burning rivers and the poisoned lands were imminent.

But much more crucially, as the novelist D. H. Lawrence was to observe, the Apocalypse of John was also infused with revolution. It was a vision of the powerful and corrupt of the world being swept away by monsters and fire and plague, and a new, cleansed world emerging for the poor, the hungry and the dispossessed. In this sense, the code of the Apocalypse was one that found repeated expression through the centuries, from France to Russia and beyond . . .

12. THE SUPERNATURAL FINGERS

Palaces of gold and marble, strange visionary dreams, fiery furnaces, dens of lions and a supernatural coded message: the Book of Daniel in the Old Testament is in some ways a beguiling code all the way through. The kings of Babylon that it depicts – Nebuchadnezzar and Belshazzar – were real, as was the destruction and plunder of Jerusalem and the captivity of the Jewish people. Nebuchadnezzar, indeed, bestrode the Middle East and the lands between the Tigris and the Euphrates, and built an empire of extraordinary wealth and splendour. Yet in the Old Testament, both he and Belshazzar are tormented by dreams and riddles that they cannot interpret. And it is Belshazzar who sees the writing upon the wall of his palace. The Old Testament codebreaker in this and other instances is the noble Jew Daniel, and the story now is an intriguing one, not merely of unlocking meaning but also how such intelligence is then handled to manipulate the course of history. It's not just what the writing on the wall, when deciphered, says. It's what happens after its message of foreboding has been digested.

The Book of Daniel is hypnotic. Using historical reality as the loosest springboard, it not only explores the relationship between God and over-mighty kings but also underscores the importance of lively wits. In Nebuchadnezzar, whose civilisation-building reign spanned much of the sixth century BC (Babylon was situated in what is now Iraq), there is a portrayal of the neurosis that comes with total power. Here is a king who can have individuals, families, villages and cities put to death upon a whim. Yet he is troubled by enigmatic dreams and cannot help voicing these fears to a retinue of magicians and astrologers he holds close to him. One dream is so terrible that he summons all of his wise occultists to interpret it for him, but the setback is that he now cannot remember the dream. The seers are set an impossible challenge, not only to interpret the dream but also to tell the king what the dream was in the first place.

They cannot. And for 'this cause, the king was angry and very furious, and commanded to destroy all the wise men of Babylon'. But Daniel can see a way through. 'Then was the secret revealed to Daniel in a night vision. Then Daniel blessed the God of Heaven.' Armed with his divine knowledge, Daniel tells the king what he dreamed. It was a great image with a head of gold, 'breast and arms of silver', thighs of brass and feet of clay. The feet are 'breaketh in pieces' and then soon after so is the rest of the statue. What can it all mean? The anxious king is desperate to hear. Why, the golden image is his kingdom, and the lesser materials below are the kingdoms that will follow.

In real life, Nebuchadnezzar's realm was one of the great wonders. Entrance to the city of Babylon was through the Ishtar Gate, a dazzling rich-blue-glazed wall emblazoned with motifs of lions. There were ziggurats and waterways, marble and tapestries, the light of gold and silver, and of course the fabled hanging gardens. Yet in the Book of Daniel, all such splendour is fleeting and Nebuchadnezzar, in all of his murderous impulsiveness, is painfully aware of the fact. After some slips, including telling his people to venerate a vast golden statue, and throwing Daniel's fellow wise men into a fiery furnace (from which, thanks to God, they emerge wholly unsinged), Nebuchadnezzar was careful to make sure that all knew to put their faith in Daniel's Lord.

But after another of his sinister dreams (mighty trees, disembodied voices) leaves him in fear, Daniel the interpreter has to tell Nebuchadnezzar the truth: he is the tree, and in time men will seek to cut him down. Indeed, Daniel foretells that the great king will be pulled down so far that, for a time, he will live among beasts. The king loses his mind and is ejected from his own sumptuous palace. For seven years, he lives on grass and dew in the wilderness, among cattle.

But Daniel's puzzle-solving services are needed further by Nebuchadnezzar's son Belshazzar (not his historical son – more likely his grandson – but the Book of Daniel is happy to conflate for the purpose of the moral lesson). Belshazzar has inherited a city that glitters with gold and wealth, and he is keen that he and his courtiers should live in the brightest splendour. Thus it is that he makes his own grave and terrible mistake. When ordering a great feast in his palace, Belshazzar uses the gold and silver goblets and plates that had been stolen in the great plunder of Jerusalem. 'They drank wine, and praised the gods of gold, and of silver, of brass, of iron, of wood and stone.' That – as they don't say in the Old Testament – does it. Heavenly retribution is swift and chilling. 'In the same hour came forth fingers of a man's hand, and wrote over against the candlestick upon the plaister [*sic*] of the wall of the king's palace: and the king saw the part of the hand that wrote.' But the king cannot make any sense of the words that have been inscribed upon the wall.

Calling in astrologers and other wise men, he declares: 'Whosoever shall read this writing, and shew me the interpretation thereof, shall be clothed with scarlet and have a chain of gold about his neck, and shall be the third ruler in the kingdom.' In other words, he who supplies the intelligence will find himself awarded a key role in the functioning of the state. In this sense, Daniel was the forerunner of the Babylonian GCHQ.

Daniel turns down all honours, insisting he will do the work without reward, before tackling the wall code that baffles Belshazzar. He then takes the opportunity to speak truth to power. He tells of Nebuchadnezzar's fall, and his years of madness, of Belshazzar, who, despite knowing all this, hardened his heart against the Lord and drank from 'the vessels of his house' with an array of 'wives and concubines'. Thence to the writing on the wall. It consists of the words ME-NE, ME-NE, TE-KEL, U-PHAR-SIN.

There is not a wise man or astrologer in Babylon who can fathom them, but for Daniel, the words resolve instantly into solid meaning. ME-NE means that God has 'numbered' Belshazzar's kingdom and 'finished it'. TE-KEL, means that he was 'weighed in the balance and found wanting' (it is always so satisfying to see the origins of familiar phrases), and the other letters signify that the 'kingdom is divided' and will fall to the Medes and the Persians. And so it comes to pass, with remarkable speed, for Belshazzar is assassinated that very night by Persians and the kingdom is taken by Darius the Medan.

At first, Daniel's immense interpretive skills are very welcome and Darius 'thought to set him over the whole realm'. Jealous courtiers have other ideas though, and they turn the king against Daniel and have him cast into the den of lions. But Darius, after a sleepless, tortured night, cannot bear it, and at dawn he makes his way to the sealed den and cries out for Daniel. The angels of Heaven have saved him and ensured, miraculously, that the lions in the den have left him be. Once again, Daniel is enfolded deep into the bosom of the state. The ending is less happy for the jealous courtiers; not only they but their wives and their children too are cast into that den of lions, where the beasts 'had mastery over them' and 'breakst all their bones'.

The story of Daniel, while confirming the supremacy of the Lord over earthly kings, is also at its core a codebreaking morality play. It is not just strategic intelligence that kings seek but the truth. There is something about the unravelling of a code – rendering its true meaning naked – that elevates it above other forms of espionage. Daniel's codebreaking of the writing on the wall told of a forthcoming coup and conquest but the intriguing part of the story is that Belshazzar, rather than reacting in anger and denial, sought to reward Daniel further for giving it to him straight. He spoke plainly of death and downfall – not all codebreakers have received such a warm response upon the delivery of bad news.

1

SHADOW CODE

In this puzzle, the code is hidden in the shadows. Put the answers to the questions horizontally in the upper grid (all answers have seven letters). Take the letters in the shaded squares and place them vertically, one below the other, in the lower grid.

When you have completed the lower grid, the shadow code will reveal a famous biblical quotation.

1 Drawing showing a plan or outline.
2 A mobile home or a desert convoy.
3 Pungent accompaniment to roast beef made from small seeds.
4 Person skilled in physical exercise especially track and field events.
5 Tanned animal hide used in making shoes, bags, etc.
6 Small smooth stones found on a beach.
7 Venue where plays are performed.
8 A thrifty purchase, a snip!
9 Coastal resort which is a holiday destination.
10 Period of time a batsman in cricket spends at the crease.
11 A strong underground cell or prison.
12 The art or process of preparing food to eat, cuisine.
13 A cross-breed of dog.
14 Black and white Antarctic bird which can swim rapidly underwater.

1							
2							
3							
4							
5							
6							
7							
8							
9							
10							
11							
12							
13							
14							

1	2	3	4	5	6	7	8	9	10	11	12	13	14

2

THE WRITING ON THE WALL

Look at the writing on the wall. Follow the instructions to remove some of the words to expose the warning in the secret coded message.

	1	2	3	4
A	LINK	WALLS	MIX	VIVID
B	OATS	REAL	ROTATOR	BEAM
C	HAVE	LEVEL	KAYAK	CLAIM
D	EATS	EARL	EARS	EACH

1. Remove words which can have the letter B put in front of them in Column 1.
2. Remove words which are made up of Roman numerals in Row A.
3. Remove any words which are made up only from letters in the first half of the alphabet.
4. Remove any palindromes.
5. Remove any words which are anagrams of each other.

What is the secret coded message?

3

BEASTLY NUMBER

Here's an addition sum with symbols taking the place of numbers. It may be a beastly puzzle but the dreaded number of the beast, represented by a 6, does NOT appear. (Iron Maiden fans are still invited to try and solve this.)

Can you crack the code and reveal the numbers involved?

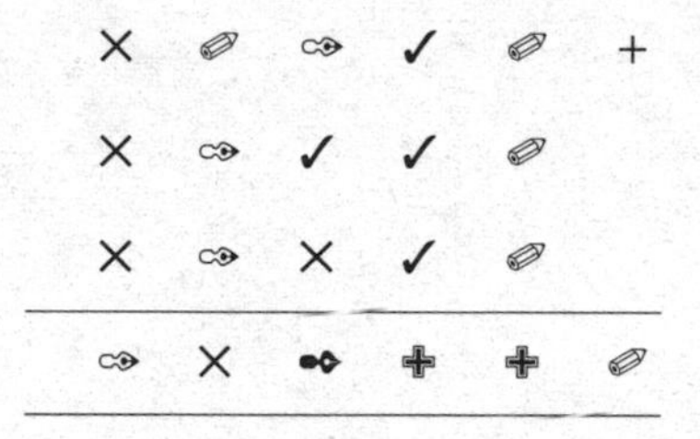

CHAPTER FOUR

PYRAMID POWER

In which we see how Egyptian myth – and the quest to decode the uncanny secrets of the Pyramids – inspired some unexpected and brilliant minds.

13. NAPOLEON COMPLEX

Amid the almost unearthly rose-red sunsets, the shimmering dunes and the stark ancient temples of ghost civilisations, Napoleon Bonaparte became intoxicated with the mysteries of the past. Here, with his armies in Egypt in 1798, he felt himself to be reaching back through time, communing with the shade of Alexander the Great. Alexander had conquered the land of the pharaohs and Bonaparte intended to do the same. But during the course of his campaign, his soldiers made a discovery that was to have a far longer-lasting legacy than any of Napoleon's plans. It was a vast grey stone, covered with etched writing, dating back many hundreds of years to the time of Ptolemy. The race to unlock its secrets became a cryptographical challenge that, when decoded, would change the way that history was understood.

The wonder of it is that that stone is now viewed by hundreds of people a day; it can be seen at London's British Museum and it has been based at that site for over two hundred years.

The Rosetta Stone is, in a way, living history. Those impossibly ancient inscriptions upon its surface are given fresh life by every new visitor from around the world who walks into Room 4 of the museum to gaze upon it. The joy of its story lies in the way that its surface gave scholars the key to unlocking a language which had died centuries before.

And the Rosetta Stone was a chance discovery too, an exceptionally fortuitous one. When setting out to seize Egypt from the ruling Mamluks – chiefly as a way of linking up with an ally in India and disrupting the growing power of the British Empire – Napoleon Bonaparte took not only soldiers with him on that voyage across the Mediterranean but a gaggle of scholars and scientists as well. In 1798, all of fashionable Paris was agog about Egypt, this ancient land of mysteries. There were all sorts of enigmas, from mathematics to engineering, that Napoleon and his intellectuals wanted answers to.

And Napoleon was also very careful to instruct his soldiers that when in that land they had to be respectful to Islamic practices and that mosques and religious customs were never to be disrupted or harassed but instead treated with reverence. And as he advanced through the old cities and ports, Napoleon also sought to tell all their inhabitants that he meant no harm to them in any way, and that he would uphold their ways of life. By contrast, he insisted, the ruling Mamluks were the ones who were insulting and sacrilegious, and the same went for the Ottomans.

This line worked for a while, helped by the fact that Napoleon also took to dressing himself more in the local fashion. And as his soldiers moved through various settlements, quelling attempted fightbacks from assorted Mamluk forces, the French scholars moved alongside, utterly fascinated by this antique realm. However, one formidable obstacle prevented them grasping its rich history. No one at that stage had any idea of how to approach or understand Ancient Egyptian hieroglyphs.

The discovery of the Rosetta Stone changed all of that. This relic, half buried in sand and soil, part of an old wall and its foundation, was discovered by French army officer Pierre Francois Xavier Bouchard, in the city of Rosetta. It was fragmented, certainly, with much of its original mass missing, but what remained of the Stone contained inscriptions beyond value. Some were in the form of hieroglyphs, but another layer was in demotic Egyptian and another in Ancient Greek. It was established relatively swiftly that the Greek and the Egyptian inscriptions spelled out the same message and that this was a decree issued at the time of Ptolemy. This would surely mean that the unfathomable hieroglyphs would also form the same decree, and make it possible to decode each and every one of them.

Perhaps appropriately, given the intense imperial tension between the two nations, the competitors to be the first to decrypt these ancient writings were an Englishman, Thomas Young, and a Frenchman, Jean-François Champollion. The Egyptian demotic script was the more straightforward language among the inscriptions, but it was not without its own knots of difficulty. And the hieroglyphs were on their own plane of complexity. For a while, it looked as though Young was making the speedier progress with both. But Champollion was versed in the old tongue of Coptic, which bore a relation to the Ancient Egyptian, and he had also fathomed that the hieroglyphs were to be spoken aloud: they were phonetic.

And it was Champollion's knowledge of Coptic that helped him to unravel the hieroglyph that meant 'to give birth'. This, to use the argot of codebreaking, could act as a sort of crib and from here, Champollion made leaping bounds of progress.

Ultimately, he became the first man in Europe to translate and to speak the language of Rameses and other long-dead pharoahs. It was akin to summoning the ghosts of the ancient past. The effort involved in doing so had the most intense physical effect upon him. After exclaiming in triumph that he had unlocked the language, Champollion fell into a profound faint. It was said that he remained in this curious coma for a week, as still and unmoving as one of the mummies that lay as yet uncovered in the old tombs. So what was the decree passed down upon that stone?

It was issued under a thirteen-year-old boy-king who reigned two centuries before the birth of Christ. His name was Ptolemy V Epiphanes. Priests had assembled in the city of Memphis, where different stones were to receive the same inscription. The reign of the previous king of Egypt had been marked by rebellions. These had been quelled and the coronation of the thirteen-year-old heir was celebrated on the Stone in a message aimed at that synod of priests. There were many such stones, all with writing upon them copied from the original, placed in all the temples of all the cities.

The decree asserted the new young king's authority, while at the same time declaring how he would protect the priests, raise taxes for their temples, give amnesties to prisoners and offer other forms of financial support for ceremonies, sacrifices and feast days. In a sense, the content of the Rosetta Stone's code was not quite as significant as the fact that – after 2,000 years – it could be read at all.

And how did it come to find a home in the British Museum in 1802? Napoleon was defeated in Egypt by the British in 1801 and the Stone became British property under the Treaty of Alexandria. Even though it now seems extraordinary that such priceless history could simply have been regarded as a prize of war, the Stone was nonetheless cared for. It was shipped to Portsmouth, and thence to the relatively new British Museum, where, owing to its weight, it had to stand on a specially strengthened section of floor.

Since those days, many very much more valuable relics have been discovered beneath the timeless sands, but the Rosetta Stone continues to exert fascination because it represents a lost world restored, in all its limitless cultural richness. This was a decryption that made history immeasurably fuller.

14. GRAVITY AND MAGIC

You might have thought that laying down the foundations for modern physics and the study of optics would have been enough for one great mind in the seventeenth century, but this gaunt young man, piercing of eye, long-haired and with a clumsy dog called Diamond, was insatiably hungry for answers to the world's mysteries. So much so that his apple-falling, epoch-changing work on gravity was not enough. Sitting in his study in Lincolnshire, Isaac Newton spent a huge amount of time brooding on the pyramids of Egypt. Partly this was because he was convinced that the Ancient Egyptians had wisdom and technology that had been lost to the ages. But it was also because he thought that the pyramids contained encoded clues to biblical prophecies and the end of the world. Not for nothing was the first of the great modern scientists also known as the last of the great sorcerers; for Newton, the two disciplines were intertwined.

Newton was obsessed with the idea of cubits, the system of measurement used by Egyptian engineers and mentioned in the Bible. His theory was that calculating the exact measurements of the ancient tombs could in turn lead him to revelations concerning a long-lost sacred structure that itself contained cosmic secrets: the Temple of Solomon that had once stood in Jerusalem. He believed close study of the pyramids' esoteric dimensions would enable him to mentally reconstruct the Temple, and in so doing delve deeper into its own coded information about the end of times and the coming apocalypse. Such matters were scarcely orthodox thinking in England in the 1680s. And Newton had other heretical beliefs (including that Jesus was an intermediary between God and man, rather than the actual son of God) that he knew were best confined to private notebooks.

That house in Lincolnshire, Woolsthorpe Manor, had its place in history assured even when Newton was a young man. Studying and then working at Trinity College Cambridge, had set him on his scientific path, but it was among the low beams of this house that his genius really sprang forth. It was here that Newton's laws of motion were formulated, with insights on gravity and thoughts on the movements of planets. It was here also that astonishing leaps in mathematics were made, with Newton honing calculus and other mind-tangling theorems. Within these walls he experimented with the refraction of light and summoned rainbows from prisms.

With a whole world of wonder being opened up by his inquiries, it was perhaps understandable that Newton's astounding imagination should set to work envisaging the construction of the pyramids in those ancient days with all the passages and galleries and tombs they contained. He examined a range of classical sources for references to 'royal cubits', intent upon using them to unlock the prophecies of Daniel and John, and also to prove that the Ancient Egyptians had formulated a means of measuring the circumference of the earth.

Some of Newton's handwritten notes on the subject were sold not that long ago at Sothebys (some were scorched, because Newton's dog Diamond had jumped up too enthusiastically and knocked a candle off a table). Newton's interest in mystical encryptions is important because such beliefs went some way towards inspiring the classical physics and mathematics that the modern world was built upon (and the same is true for his obsession with alchemy – turning base matter into gold – from which also came more fruitful avenues of scientific genius for him to explore). And across hundreds of thousands of words inscribed carefully in countless notebooks, which disappeared for centuries while being held in private libraries, Isaac Newton also employed ciphers. Sometimes they were an elaborate form of shorthand used by other figures in the 1600s, and other times he deployed transposition codes. Partly these were to disguise confessions or outbreaks of heresy, but on other occasions they were used simply to prove that he had had an idea or an inspiration first. Whenever a new discovery, whether in maths or physics or optics, was proclaimed by European rivals, Newton's coded messages could then be unravelled. Decoded, they would read as the exact discovery or inspiration already articulated by Newton.

Throughout the course of his long and extraordinary life – he was made president of the Royal Society, and knighted – Newton's passion for esoteric knowledge continued to flare and spark like fireworks. One might have thought that discerning and describing the moon's gravitational power over the tides would have been magic enough, but Newton's quest to burrow deeper into the darkness of the pyramids illuminated how in the 1600s the compulsion to understand the natural world fused empirical thinking with the supernatural. This philosophy was the inspiration for Newton's very real and world-changing science and mathematics.

Since the 1600s, those pyramids and that lost temple have continued to exert a codebreaking fascination, with swarms of others rather less gifted than Newton seeking to decrypt their structures. One popular New Age theory was that the cipher of the pyramids, once unlocked, would disclose the dimensions of other planets and suns. Unlike Newton, who ceaselessly sought empirical evidence, this branch of cryptology is – as has been gently pointed out – more to do with staring at mathematical structures and overlaying patterns on them.

1

PYRAMID BUILDING

Hidden meanings lie at the heart of the fascination with the pyramids. This pyramid shape is built up using 27 words. Can you match the words into pairs which have very similar meanings. There is one word which is all alone and does not have a match. What is it?

ASK

STUDY

COLLECT

TRY ALTER

METHOD WORD

UNRAVEL CHIEF

DISCOVER CHANGE

ACCUMULATE CIPHER

QUESTION EXPRESSION

CODE LABOUR TERM FIND

WORK MAIN FACTS ATTEMPT

TRUTH SYSTEM PHRASE SOLVE

2

POINTED

The pyramids are certainly pointed in shape, but did the shape and size of them point to some greater mystery?

Rose-Ella Sloane is relaxing in her Kensington flat after a day researching at the British Museum. Like Sir Isaac Newton before her, she is preoccupied with the secrets of the stones. Her doodles and diagrams have led her to produce a puzzle. There are TWELVE words each containing FOUR letters. She writes them in the mini pyramids with each starting letter fixed in a numbered area. The second letter goes directly above the first, the third letter goes to the right of the starter letter, and the fourth letter goes to the left of the starter letter. She discovers that there is only one way for all the words to interlink in this pattern. When both pyramids are completed, the letters reading across in the numbered areas on the bottom row reveal a word which would have been very familiar to Sir Isaac. The sands of time may well drift by as you try to solve this!

CLUE: The letter in triangle 12 is in the second half of the alphabet.

ALSO	BLUE	CART	CODE	FOIL	IDEA
ITEM	OILS	SAVE	TREE	USED	YEAR

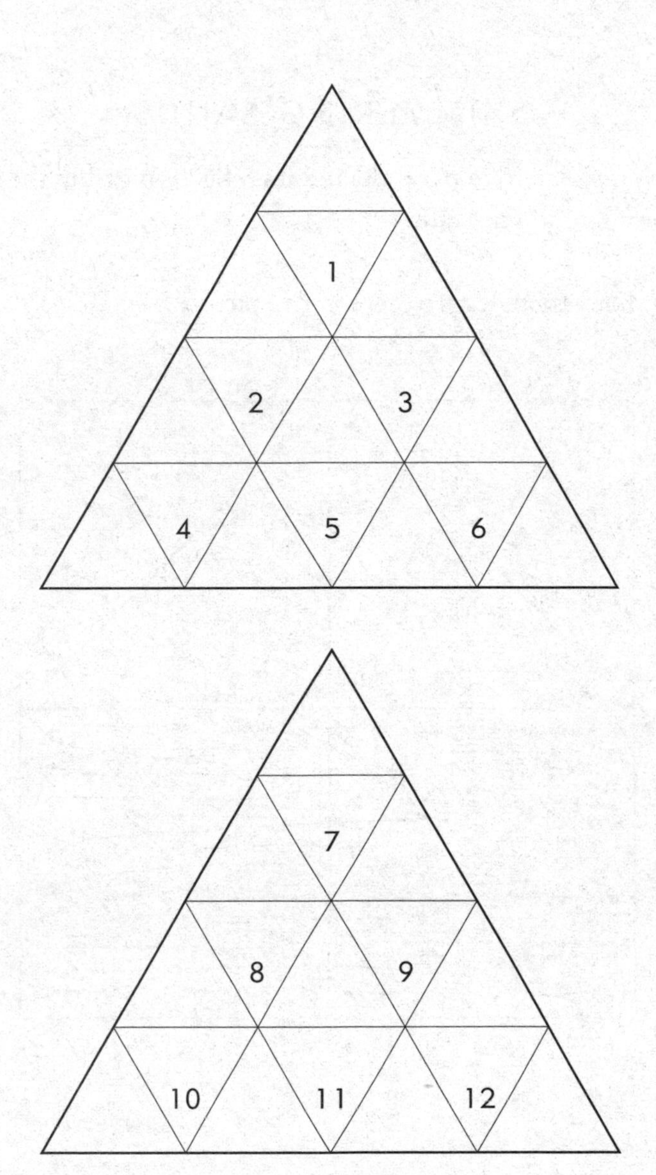
1
2
3
4
5
6
7
8
9
10
11
12

3

SHIMMERING SANDS

Here's a picture of a pyramid and its reflection cast in the heat of the shimmering sands.

How many triangles are there in the picture?

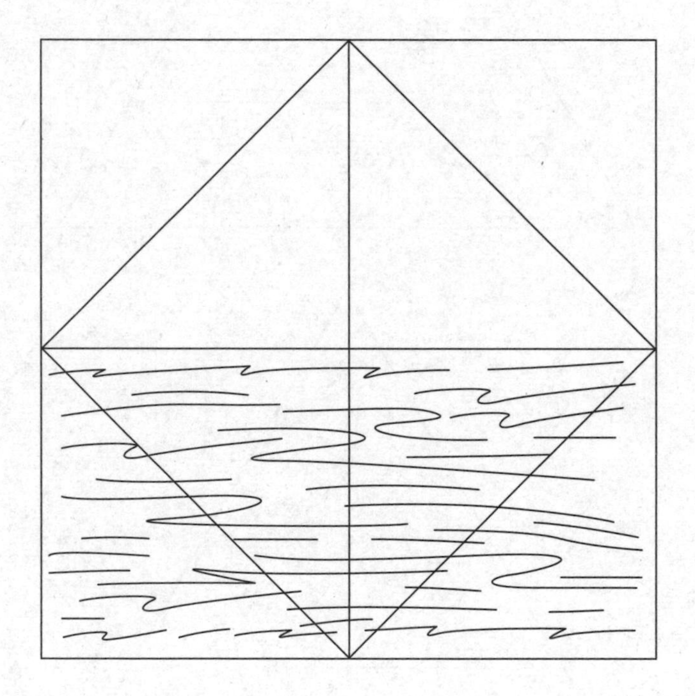

4

PATTERNS

The very dimensions of the magnificent pyramids may be a coded device holding some ancient mystery. To the modern eye, there is definitely a pattern that runs through the letters and numbers appearing in these groups of pyramids. In each case, what should take the place of the question mark?

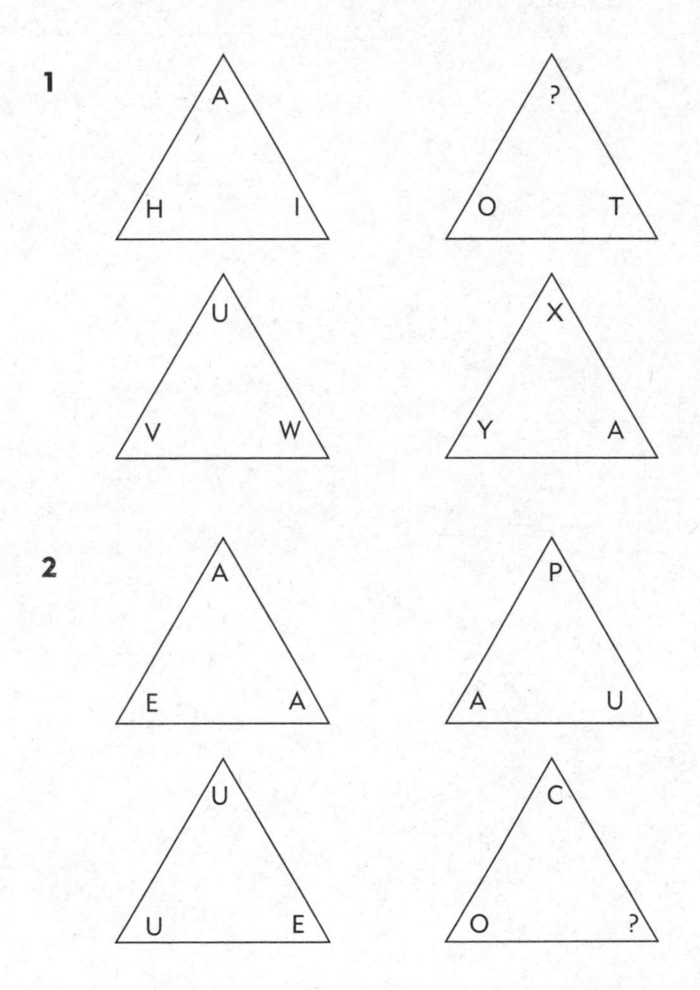

CHAPTER FIVE

THE SECRET CIPHERS OF LOVE

In which we examine how codes and ciphers have through history always been linked with the ingenious stratagems of romance.

15. THE INDIAN LOVE SONGS

This was – and is – a vast and mighty land of rich cities, extraordinary art, endless fertile plains and mountains that fill the sky. India is the seventh largest country in the world by area, and carries an intense, intricate and complex history. Over two thousand years ago, India was stunning travellers, traders and would-be conquerors with its architecture, its carvings, shrines, solemn tombs – and its literature. One of the lasting works that this period has given the world is the Kama Sutra. It is often used as a byword for sexual inventiveness and the celebration of sensuality – a textbook of positions, to put it crudely! But it also offered a great deal of wisdom – including teaching women the art of rendering their writing secret, so that their assignations might remain discreet.

Thought to have been written around the third century CE, the Kama Sutra was the work of a man called Vatsyayana, and his name is all that is known of him. But the book, in Sanskrit, contains a great deal of intriguing detail about society and sexual politics in that far-distant era. The man and the woman who are followed throughout the text are taken through stages of their lives, and they have associates and friends. As well as the famously detailed descriptions of a variety of love-making techniques, there was much here too about power relations between men and women. 'It is a book about the art of living,' one academic recently declared. The text addresses virginity, seduction and marriage, and it also dives into adultery, separation and prostitution. There are even passages concerning the 'third nature', in which illicit romance is found between gay people.

In the Kama Sutra, the arts of loving – and of preserving that love – are themselves a form of code. But the code-writing itself is a fantastic glimpse into an ancient society that understood very well the need for secrecy in affairs of the heart. The encipherment technique is called *mlecchita vikalpa* and it is there in the first volume. There were already cipher systems throughout the mighty continent, one being the secret language employed by kings and Silk Road traders alike. But the women in the Kama Sutra were also advised to adopt codes (among other accomplishments), and in particular, it is thought a specialised method known as Kautiliya. This was a substitution cipher. On a grid of two horizontal rows, consonants were swapped with vowels and vice versa. This was 'the art of understanding writing in cipher, and the writing of words in a peculiar way'.

Vatsyayana was not addressing a sensationally feminist society – very far from it. Men were regarded as active and aggressive, women as docile and weak. Yet the code-writing suggested subtler layers of psychology between women and men. Elsewhere in the Kama Sutra, women were also advised to learn other arts, such as cooking and massage. But the codes gave them a secret voice all of their own. Such writing might be employed by women who were keen to establish relationships that might otherwise be forbidden. The codes could also be used by sophisticated courtesans, giving another level of meaning to their discourses with rich and powerful men.

In this way women might also be able to correspond with friends about their forbidden encounters or indeed keep secret diaries. Code-writing is in part about people gaining a little power and agency in a world that can be freighted against them. It also signifies a hidden knowledge, and as such became part of the Kama Sutra's wider pattern of acquiring sensual wisdom. The intoxicating world that Vatsyayana summoned (there is even some transitioning; in a story where a man accidentally touches the god Shiva his punishment is to be turned into a woman, and the woman subsequently asks the god to be turned back again) is one where all the senses, as well as the intellect, are engaged in deciphering the desires and the intentions of others. His codes were part of the secret language of love.

16. CASANOVA'S SEDUCTIVE CONUNDRUMS

Only a few people are immortalised by a single name; even fewer are immortalised for notorious sexual exploits. Yet Casanova, the 'eighteenth-century Venetian gambler, eroticist and spy' as he was described on a Britpop album by Neil Hannon in a sublime 1996 Divine Comedy album (also called *Casanova*), remains a byword for licentious adventures. This is largely because early readers of his titanic autobiography cherry-picked the franker passages and discarded much of the texture and detail of the extraordinary life he lived outside of various boudoirs.

Giacomo Casanova, born in Venice in 1725, certainly led the most colourful and vivid life – as well as the bed-hopping, he indulged in travel and art and espionage. There was also, in one of the more curious corners of his life, codebreaking.

Casanova's parents were actors and eighteenth-century Venice was a sumptuous city of performance. The other-worldly quality of palazzos seemingly floating amid the canals and lagoons was transformed on sultry summer evenings into glittering tableaux of reflected flame and gorgeously coloured lanterns, a world through which sophisticated prostitutes and gamblers from all over Europe moved. Young Casanova suffered separation and loss. His father died when he was small and his mother was obliged to tour with the theatre, so he was sent to his grandmother and then placed in the care of a senior cleric. Casanova was bright; he learned the violin and was also enrolled into the University of Padua aged just twelve. Here he became something of a polymath, studying medicine, mathematics and law. But there was also sexual awakening (with two sisters) and a form of intellectual impetuosity that saw him become a gambler, join the army and then – with brilliant good fortune – find a rich patron in the form of a Venetian grandee called Bragadin.

This enabled the handsome young rake to live a life of romantic dissipation, in the sumptuous attire of a noble. His scandalous exploits aroused the hostility of the Venetian authorities and he was only ever a few steps away from being locked in a prison cell. In between fragrant bedchambers and gambling dens, legal work for Bragadin and imprisonment in the Doge's palace, from which he amazingly escaped, Casanova must have acquired a certain skill in cryptology. It was a few years later, when his adventures (sometimes as a spy) had wafted him from Venice to Dresden to Vienna to Paris, that that skill was to find the most baroque expression.

In Paris he moved in the gilded circles of Madame de Pompadour and the sometime occultist Cagliostro, and there he caught the intrigued attentions of the very wealthy Madame Jeanne d'Urfé, who was around twenty years his senior. Paris in the mid eighteenth century was a claustrophobic maze of tall houses and dark streets. Madame d'Urfé, like many at the time, was an avid believer in alchemy (turning ordinary metal into gold) and other occult pastimes, such as numerology. Once, the grand lady had written an encrypted letter and asked Casanova to deliver it, confident he would not be able to fathom the contents, but Casanova confounded her, for he had a fast, sharp eye.

It appears he was naturally predisposed to codebreaking as it took him only a matter of days to unscramble the code. Casanova looked at the frequencies of certain encoded letters and the lengths of certain encoded words, but the chief means by which he unlocked it was by guessing the word that acted as the key to the cipher. More importantly, he shrewdly guessed that the letter was actually a continuation of that lady's overriding obsession: the nonsense of obtaining gold from dross. Once decrypted, he flourished his achievement before Madame d'Urfé, who was astonished that he had managed to divine the keyword, which she and she alone held in her imagination. The keyword was 'Nebuchadnezzar', the Old Testament king (from page 71) who had been troubled by prophetic dreams.

This cryptological success brought out Casanova's more cynical side. He recognised he could use this as a lever to extract anything he wanted from Madame d'Urfé if he painted it as a mystical gift that he possessed. 'I could have told her the truth,' he wrote in his confessional memoir, 'that the same calculation which had served me for deciphering the manuscript had enabled me to learn the word – but on a caprice it struck me to tell her that a genie had revealed it to me. This false disclosure fettered Madame d'Urfé to me. That day I became the master of her soul and I abused my power.'

What the unfortunately credulous Madame d'Urfé could not have known was that such a gift would have been, for a spy, part of a more general education. The febrile days of pre-revolutionary France saw a rising need for keeping sensitive communications secret. The Foreign Ministry had devised a system of *petits chiffres* (or 'little codes') using a numbering system as a key (of which it was possible Casanova was aware). A number would be assigned to a word; numbers 1 to 600 covered 600 words. In terms of politics and war, this was deemed sufficient to cover all secret ground. By 1750, it was no longer enough and the number of words increased to 1,200. But come the days and years following the 1789 revolution, when Napoleon was trampling across the map of Europe and placed his brother Joseph upon the Portuguese throne, it was understood that even this was too easy to crack. The accursed British, for instance, seemed to have a knack for unravelling messages. The result in 1811 (some way after Casanova's time, though he would have understood the principles) was a new-look revised cipher: the Great Paris Cipher.

Here was an awesome prospect: 1,400 numbers assigned to words, with extra dummy hieroglyphs added in order to sow confusion and disrupt any decryption effort. As the Peninsular War raged, the British selected their own dedicated codebreaker: George Scovell, who had himself designed codes for the British Army. Those communications to and from Napoleon that were intercepted were sent on to Scovell. The Great Paris Cipher was indeed a challenge, but Scovell, himself a smooth linguist, managed to put together a team of Spanish- and Portuguese-speakers, among others, to tackle the job. He spent months devoting himself to breaking open Napoleon's codes. A letter from Joseph Bonaparte to Napoleon, snatched in 1812, gave Scovell the leverage he needed to throw it all wide open. He had been shrewd in making a close study of the context of uncoded communications, which gave him an idea of the subjects and terminology that might be frequently used in the coded letters. And it was with this missive, which contained a vast amount of intelligence about troop movements and plans, that the Great Paris Cipher was mastered.

Adventurers such as the young Casanova would have had no difficulty in recognising the origins of this elegant French system, and he would have also understood the idea that when it came to both war and love, sometimes it was necessary to weave webs of deception. From alchemy to battlefields, the European approach to codes was both dramatic and domestic.

17. A SYMPHONY OF GENIUS

In the 1938 Hitchcock comedy-thriller *The Lady Vanishes*, the core espionage plot involves getting a secret code across central Europe and back to the Foreign Office in London before war breaks out. Is it written on paper? No, far too vulnerable to capture. Does it even come in the form of words? Again, no, for words can in the end be decoded by the enemy. Instead – and here is a SERIOUS SPOILER ALERT, so if you have not seen the film, skip the next few words RIGHT NOW – the code takes the form of a whistled or sung melody. The unwitting hero and heroine, staying in a charming hotel at the start of the adventure, are puzzled and horrified when a folk musician serenading a very catchy tune beneath another guest's window is subsequently murdered . . .

But the conceit – very entertainingly purveyed in a still fresh and funny film – has also, for very many years (indeed centuries) tantalised real-life code-setters and breakers. Is it possible to convey secret messages through music? Composers such as Bach, Brahms and Schumann certainly thought so, and their efforts can still be heard today. But it was as the nineteeth century gave way to the twentieth that another composer, Edward Elgar, not only smuggled a message to a young woman called Dora into his music but also created a code that resists analysis even now.

The establishment of the written musical scale, with notes from A to G, always presented the intriguing intellectual and aesthetic possibility of a limited sort of message being conveyed in a melody. Composers sometimes embedded their own names as a flourish: Bach being an example (the H was achieved because in German terms it was the equivalent of a B sharp). The resulting B-A-C-H melody turned up in *The Art of the Fugue.* One would have to be a musician to recognise the notes and appreciate the light-hearted self-reference, but the fact is that there is an odd compatibility between the minds of the musically talented and the minds of codebreakers, as we shall see a little later.

In the mid nineteenth century there was a great but ultimately tragic romance between Johannes Brahms and a singer named Agathe von Siebold. He was instantly smitten with the soprano when they met through a mutual friend, and he was swift to ask for her hand in marriage. But his confidence and self-belief were shattered when his Piano Concerto No. 1 was ill-received, and he got it into his head that he would never be able to make a decent enough living for them both. And because of this, he impetuously called the wedding off. But Agathe continued to haunt his imagination and dreams. She was the true love he had cast aside. She went off to become a governess and later married someone else. Brahms, in the meantime, infused his String Sextet No. 2 with a melody based upon the notes A-G-A-H-E (once again, the H was a B sharp).

According to music historian Suzy Klein, such codes also held a particular political significance at times of great oppression. An example can be found in Stalin's Russia, where the great composer Dmitri Shostakovich used the initial of his forename and several letters from his surname as a short recurring leitmotif of D-S-(e-flat)-C-H (b-natural). The composer frequently came under terrific political pressure, living as he did under a tyrannical ruler who was responsible for countless deaths. Klein speculated that Shostakovich used cryptography to embed his name in his music to demonstrate that he could not be obliterated and would live on through his work. This was at a time when any form of artistic subversion in the Soviet Union could have the most serious consequences for careers and lives.

But could musical codes have any diplomatic or military use? In some historical cases, the possibilities were fully explored. The encryption department of the Vatican in the 1600s experimented not only with sending messages through musical notes but also widening the range of potential letters and words that could be incorporated with different musical pitches. There were also instances when small walled cities, facing sieges, starvation and fire, could communicate with soldiers beyond the wall via codes sent using church bells rung in a very particular order. Again, the range of detailed possibilities in vocabulary was narrow, but certain sequences of tolling could indicate varying levels of stress or indeed be used to issue warnings.

As the art of music evolved, so too did the ideas to use that music to convey secret messages. The symbols used in sheet music, all the crotchets and quavers and semibreves, found a double use as pictograms, standing in for letters or numbers, so that the confidential communication might be successfully disguised as sheet music for an orchestra. Within all those bars and notes lay invisible dimensions. This practice found many ingenious outlets, sometimes in the realm of crime. The New York police spent some time trying to fathom the link between written melodies and a wave of illegal bookmaking. Again, the notes stood in for letters, and the tunes (which were more jarringly atonal than catchy) conveyed those letters.

But the idea that music could be a code found perhaps its most thrilling fictional form in the sequence of notes played to greet the aliens in Steven Spielberg's *Close Encounters of the Third Kind* (1978). Obviously, when atop a mountain in the wilderness in the dead of night, ordinary language could hardly be used to communicate with visiting extra-terrestrials. And the five-note sequence – G-A-F-F (an octave lower)-C – became for a time one of the world's most famous musical phrases. In the film, the music had been composed as one of the means of signalling to the alien craft that humanity was keen to commune with the aliens. The alien craft then respond with their own musical notes, and the two melodies – human and alien – conjoin. It is arresting and strange and moving. In real life, the music was the work of composer John Williams, who made that five-note sequence the core of the entire film score. So inspirational was it that for several years afterwards, real-life UFO spotters, gathering on night-time hills across America and Europe, sang the five-note sequence up to the heavens in the hope of attracting a real mothership.

Yet the most famous of all the code-setting composers was Edward Elgar. His works were infused with mystery. He told the world that a 'dark matter' lay behind The *Enigma Variations* and that there was another piece of existing music which – if played alongside The *Enigma Variations* simultaneously – would fit it perfectly, the two seamlessly intertwining. This other piece of music was apparently famous, but what was it? Elgar died before letting anyone know. One recent theory has it that the *Enigma* itself is a counterpoint – that is, it is not the main melody but rather the accompaniment to the absent music. Over the years, all sorts of candidates were put forward, including 'Rule Britannia'. It did not fit.

Because Elgar was such a fan of cryptography, a variety of other fiendishly clever theories were tried out, including old Protestant hymns and even 'Twinkle, Twinkle, Little Star'. More recently, a young composer called Ed Newton-Rex, alumnus of the Choir of King's College, Cambridge, presented his own candidate: a choral piece by Pergolasi called *Stabat Mater.* But whatever the solution, the fundamental point is that Elgar adored mysteries and codes. So much so that a written cipher of his own devising has still yet to be satisfactorily decoded. It is widely known as the Dorabella Cipher. The composer addressed it to the young woman who also inspired Variation 10 of The *Enigma Variations* (a piece also entitled 'Dorabella'). Her name was Dora Penny and the code that was invented especially for her is still the subject of head-scratching cipher-busting competitions.

She was the twenty-three-year-old daughter of a vicar whose wife was friends with Elgar and his wife, Alice. In 1897, the forty-year-old Elgar had yet to receive recognition and fame; at that stage, he was a music teacher who composed. Within the next several years, that would all change very dramatically, and as a souvenir of this fame, Dora Penny would hold on fondly to the mysterious cipher that Elgar had sent her, eventually allowing it to be published. The circumstances, which might these days cause raised eyebrows, appeared to be less noteworthy back then: Mr and Mrs Elgar were invited to come and stay at the Wolverhampton vicarage of the Reverend Alfred Penny and his wife and daughter. This was in the summer of 1897 and the sojourn was apparently very pleasant for all parties. Upon their return home to Great Malvern, in the shadow of those glorious hills, the Elgars began writing their thankyous. That of Alice Elgar was straightforward, but that of her husband, slipped in alongside it, was extraordinary. Appearing to be addressed largely to young Dora, it was a great mass of semicircular squiggles, like rounded Ws, but arranged in all sorts of different angles and configurations.

The composer had learned about secret writing via a series of articles in *Pall Mall* magazine, but this was no simple transposition or polyalphabetical code. It looked and felt more alien than that. There was later speculation that at the vicarage there had been some discussion of the ancient hieroglyphs to be found in faraway tropical locations, and that these marks of Elgar's were a sort of homage to those glyphs. The friendship between the composer and the young woman continued, and with it came later speculation about its precise nature. Was this a clandestine romance? Was the coded letter a secret love note?

In the 1930s, Dora Penny wrote a book about Elgar – the very opposite of a 'kiss and tell', in that there was no kissing and no telling – but from that point onwards, the Dorabella Cipher seized the imagination of Elgar's musical fans. Various ingenious solutions have been proposed across the decades; the squiggles, amounting to some eighty-eight characters, have yielded plaintext suggestions, such as (in part) 'I own the dark, makes E. E. sigh when you are too long gone' and (again in part) 'why am I very sad, Belle. I sag as we see roses.'

But Elgar was also influential (not just musically) because the range of his imagination, in his compositions and ciphers, opened up and continued to hold open the promise of more complex encrypted melodies. He saw the possibilities of entire concertos that are not only musically thrilling but which also carry encoded messages that presumably these days could be unravelled instantly by means of computer. Recently a website called Clarallel explored the principle: put the letters of your name into the field and they will be transposed into musical notes, which the computer will play. The name 'Sinclair' produces a pleasingly quizzical tune. In essence, those Hitchcockian spies who made off with the vanishing lady in 1938 might have been a foreshadowing of brilliant encryption techniques to come.

And as a musical postscript, it became noted with interest at Bletchley Park during the Second World War that there seemed to be a curiously strong link between codebreaking aptitude and a predilection for music. From the composer James Bernard – a protégé of Benjamin Britten – to BBC orchestra conductors, and a range of madrigal singers and piano players in between, the Park's personnel teemed with melodic talent. It has been suggested that there is a strong relationship between the capacity of mathematical thought and the complex structures of music. Elgar might have agreed that there certainly seemed to be an affinity between harmony and the hand-woven intricacies of ciphers.

18. THE CODE THAT DARED TO SPEAK ITS NAME

In the 1960s, there was a BBC Radio comedy series, *Round the Horne*, which featured regular sketches involving two gentlemen called Julian and Sandy. They were resting actors who, in inventive entrepreneurial style, were always taking on other work when the phone was not ringing. Whether becoming bookshop managers, all-in wrestlers, travel agents or musical impresarios, the other key point about this hugely loved duo (played with wild, camp relish by Hugh Paddick and Kenneth Williams) was their extraordinary use of language. In each episode, they would greet Kenneth Horne with the words: 'How nice to vada your dolly old eek! What brings you trolling in here?' (Translation: 'How nice to see your charming face again – what's the reason for your visit?')

It was the language for those who loved in a way that dared not speak its name. Julian and Sandy were using Polari, an elaborate and rich slang system that was thought to have arrived in Britain via sailors coming into ports. By the twentieth century it was being used by gay men at a time when homosexual acts were illegal. It was a means of subtly signalling one's orientation without attracting the hostile attention of the authorities. 'Vada the lallies on that bona omi', for instance, meant 'look at the legs of that nice gentleman'. 'Nish the chat' meant 'stop talking', 'naff' meant low quality, 'dolly palone' was an attractive woman, 'eek' was face and 'lattie' was house.

There were theories that Polari had in part originated in Italy and had migrated across Europe and that its use among aforementioned naval crews then spread when they came ashore. It gained particular traction in Britain in the nineteenth century with the rise of the music hall and of an entire theatrical tradition. It also spread throughout travelling fairgrounds, the clandestine language of showmen. But it found most eager use among gay men and women in an age when their very existence was considered a moral outrage. Here was a secret language that signalled and conferred amused complicity, but it also suggested that the speakers were in safe, trusted company. Polari had layers of humour, but its purpose was serious for any who might otherwise have felt isolated. To speak and understand it was to be part of a community that was underground but supportive.

In a curious way, it also conferred a sense of equality, when used by women and men in the theatre. It was a code that suggested a level of unshockability and insouciance. In this world, dancers were 'wallopers' and clothes and costumes were 'drag'. There were certain London pubs where Polari was the dominant form of discourse. And the use of Polari widened out in the mid twentieth century, back out to sea and the thriving merchant navy. Many gay men went to work on cruise ships filled with omis and dolly palones, some of whom were dead naff. Eventually, the mid-1960s incarnation of Julian and Sandy, whose every appearance on *Round the Horne* was greeted with gleeful audience applause, brought Polari into the mainstream. Even the strait-laced Kenneth Horne began acquiring it, as in the episode where Julian and Sandy have become directors of art-house films. 'Would I have vada'd any of them?' asks Horne. Julian is astonished at Horne's command of Polari and Sandy wonders 'where he spends his evenings'.

But this comic exposure also ended the exclusivity of Polari – now everyone could imitate the secret code. In 1967, homosexuality was decriminalised in Britain and a new generation wanted to emerge from the suffocating shadows. Why should gay people have to use secret codes in this new world? And so it was that Polari's use slowly began to fade.

There have been outbreaks of nostalgia though, for this, an erstwhile secret language of forbidden love, because it is possible to make a case that it made a real difference to the lives of those who spoke it. At a time of criminalisation and repression – not to mention violence and prejudice – Polari was a code that signalled solidarity among marginalised people. And the wider point about characters such as Julian and Sandy was that the audience always sided with them, and indeed loved them. The naughtier the Polari, the more the audience roared. As depictions of gay men, they were, for the time, rather positive and enduringly cheerful (when serious dramas were by and large portraying gay men as haunted blackmail victims or trapped in unhappy heterosexual marriages).

To a certain extent, many social groups form their own language – cockney rhyming slang is another widely lauded example – but Polari had a particular resonance because of its curious richness. The first use of the term 'bona' actually materialised in Shakespeare as a line in *Henry VI*. And the word 'naff', which acquired wider currency meaning 'poor quality' and 'tasteless', was thought to have Italian roots stretching back centuries. But on top of all that, Polari had an enthusiastic musicality, as when Julian and Sandy exhorted the bald Kenneth Horne to let his 'riah' (hair) right down.

1

LOVE'S OLD STORY

Love songs have been passed down through the mists of time as a means of expressing affection in a beautiful way.

This code has certain similarities to the ancient Indian substitution cipher known as the Kautiliya. Four lines of a twentieth-century love song have been written in code. Can you decipher the words and relive love's old story?

L V I T E R A E T H N – O E S H G E T S T I G

T E L E T E T E A E T H N – H O D S Y T H L T S T I G

I N Y O E H T A E A B I G – O L H P T A F T M Y R N

L V S T R T Y U – O E S O Y O O

2

CASANOVA'S KEYS

Casanova is notorious for his amorous encounters so you might be thinking that these code puzzles refer to his ability to enter, very secretly, the bedrooms and boudoirs of his paramours. However, Casanova was also a very expert perpetrator of espionage and decipherer of codes. He was adept at pinpointing keywords, which indicated which cipher was to be used.

In this puzzle, the word KEY appears in every answer. By imitating Casanova's craft and cunning, can you crack the cryptic clues and unlock the mysteries of the keywords?

1 Spirit of the Emerald Isle
2 Makes an ass of himself
3 Go ape
4 First reposition alarm; nonsense!
5 It's game but doesn't hail from where you think it might
6 Central principle of a US film studio of the silent era
7 Outwitting from the back of a horse?
8 Slow music off the Florida coast? (two words)
9 Just fine (two words, two keys)
10 Shake a leg; and an arm; go full circle (two words two keys)

3
THE LOOK OF LOVE?

There can often be more than one way of viewing things. Is it always true that the image you perceive is the correct and only visual interpretation that can be made?

This is one of the most famous illusion puzzles of all time. What are you looking at? It's not a simple case of black and white!

4

PERSUASION ABBEY

Down the years, star-crossed lovers, for a whole host of reasons, have found ways to send messages which are only decipherable to the object of their affections.

Here is a letter written by Emma Dashwood in the nineteeth century on her return to Persuasion Abbey, set deep in the beautiful Hampshire countryside. Her message to her cousin, the dashing Captain Henry Darcy, contains a hidden message. Can you decipher it?

To return to Persuasion Abbey has always been my fondest dream. My dearest cousin, we have not enjoyed each other's company for a long time. It has been too long, I declare.

Will you be guided by my judgement?

The Italian garden is now far too shady and drab. It is in danger of getting more and more overgrown day by day. Surely the time has come to let in the sunlight? The effect would be felt by every plant and bush. Even thinking of the beautiful blooms makes me smile.

Why not start this very day?

Do you not recall the dainty flowers of the brightest blue? Surely you remember the path where the roses once grew? Close to the tall cypress there was a lovely stone seat. A tiny stream ran by in which the water was crystal clear.

This most magical of gardens was a splendid sight to see.

I have such fond memories of this place where I spent many happy hours with cousin Walter, before he became a curate.

5

ELGAR ARRANGEMENT

What was the secret of the Dorabella Cipher and did Edward Elgar weave other coded messages into his works? This challenge is to look at a new Edward Elgar arrangement. The seven different letters in his name are presented as symbols. There is one shape for each letter and that stays constant all the time. As befits a composer with a mind for coded arrangements, in each pair the first word is spelled backwards to form the second word.

1 ✤ ✣ ✷   ✷ ✣ ✤

2 ✲ ✡ ✷   ✷ ✡ ✲

3 ✤ ✣ ✣ ✲   ✲ ✣ ✣ ✤

4 ✷ ✡ ✲ ✤   ✤ ✲ ✡ ✷

5 ✭ ✡ ✧ ✣ ✲   ✲ ✣ ✧ ✡ ✭

6 ✤ ✲ ✡ ✷ ✣ ✲   ✲ ✣ ✷ ✡ ✲ ✤

6

SECRET SONG

A music stave has five lines and four spaces. The lines represent the notes E, G, B, D, F in ascending order. The spaces, again in ascending order, represent the notes F, A, C, E.

Bashful Bertie could not bring himself to utter words to the woman who was the love of his life. He had to use what he knew best, the notes and the lyrics of a song. In creating the code, however, he was very sharp. He didn't want his message to fall flat. What were the words of Bertie's secret song and what was its title?

O _ _   T O   _ R _ _ _

_ _ _ R   _ R _ _ _

_ _   N O T   _ _ R _ I _

_ _ _ _   _ O O _   _ N _   _ _ _

T O _ _ T H _ R   _ O R   _ _ _ _ _ _ S

7

REBUS

A rebus is a code mixing pictures, sounds of words and letters. Here's a request from a young man and the reply from the object of his affections. The individual word clues are mixed up and not in order. The decoded messages have words of the following number of letters: HIS REQUEST 3.1.3.3.7.10? HER REPLY 1.2.3.3.3.4!

HIS REQUEST

CHAPTER SIX

THE AXEMAN'S ENIGMAS

In which the violent and bloody turmoil of the Tudor court is viewed through the prism of its deadly fascination with ciphers.

19. THE QUEEN'S BEER BARRELS

Her surrounds were sumptuous, but they were also a prison. She wore the finest furs and silks, but her poor body was wracked with rheumatoid arthritis. She was a deposed queen, yet she was not wholly powerless. And her use of secret codes was instrumental in creating the modern shape of the intelligence services. Mary, Queen of Scots now seems in some ways like a figure from a romantic historical novel: a proud woman who was also a suspected murderer, a monarch torn from her throne and a devoted lover. But in the late sixteenth century, her story and her fate had deep implications far beyond the borders of England, across a continent about to flow with blood over religion.

'The theatre of the whole world is wider than the kingdom of England,' Mary declared. Her arch-enemy and cousin Queen Elizabeth I had good reason to fear that those tides of blood would lap around her. The secret ciphers that Mary used concerned a plot to assassinate Elizabeth and would later condemn the Scottish queen to a death beneath the executioner's axe. This was not just about rivalry for the throne of England, this was also about how the people of England would pray. The Protestant revolution was still young and there were vast numbers of Catholics across Europe who wanted to stamp it out. Secret plans were being discussed for a Spanish invasion that would violently eject Elizabeth from the throne and replace her with Mary.

The affair of the secret ciphers, known as the Babington Plot, was brought to light by Elizabeth's spies in 1586. By that time, forty-four-year-old Mary had been a prisoner for nineteen years. Her prisons were (apart from the ruined Tutbury Castle, where she suffered in a freezing dark room) well appointed, largely castles and country homes belonging to trusted barons, and she had a retinue of servants and her belongings. But her supporters yearned for her release and for her ascension to greater power. The counter-espionage operation against her involved spies reaching deep into Europe. Upon those encoded communications between Mary and Catholic gentleman Anthony Babington hinged the spiritual soul of the nation. But how had it come to be necessary that their explosive letters be turned into ciphers? And equally, who were the pioneer codebreakers who unlocked their meaning and then diabolically used the same codes to gull the writers into hastening their own dooms?

Mary, Queen of Scots was born in 1542. Her world from infancy was wrought with intrigues and politics. Her father, James V of Scotland, died when she was six days old and it was then that baby Mary inherited the Scottish throne. Henry VIII wanted a match made between infants, proposing that when she reached the age of ten, Mary would be married to his son Edward. Yet when Henry's forces mounted a series of incursions into Scotland, Mary's fate was realigned and a union with France meant instead a future marriage to the (then) three-year-old Dauphin Francis. The child Mary had to be spirited across the sea. But her childhood in the French court brought her a rich education, from music to languages. Marriage at last came when she was aged fifteen; as well as becoming Queen Consort of France, her own Scottish throne was promised to the French should she die without issue.

The seeds of her doom were sewn in the wake of the death of Henry VIII's daughter, also Mary (Tudor). Elizabeth acceded to the English throne but the French (and a few in England) thought that, for reasons of legitimacy, the English crown should rightfully pass to Mary and her husband. Thus Elizabeth and Mary were fixed as lethal rivals. And the violent shocks to Mary's life were to continue. Her young husband died of an infection, and as an eighteen-year-old widow, she returned to Scotland to take up what was now an unfamiliar throne in an unfamiliar realm. She was Catholic; her nation was not. But despite the previous decades of hideous religious conflict and persecution in England – Catholics and Protestants alike burned alive at hellish stakes – she was tolerant of the upstart faith. By the age of twenty-two she had remarried, now to her cousin Lord Darnley (who also had his own claim, through the line of his aunt, upon the English throne) and the following year she gave birth to their son, James. More violence was to follow when Darnley was murdered in murky circumstances. Mary married once more, this time to the Earl of Bothwell, but Bothwell was among those suspected of having murdered Darnley. In this feverish atmosphere, the Scottish court rose against Mary, forcing her to flee to England. But her cousin Elizabeth could never risk her roaming free. Among her many prisons was the Derbyshire town of Buxton.

At this point in Elizabeth's reign, the air seethed with plots. A scheme to get rid of her, involving Philip II of Spain and the Catholic League in France, drew a young Catholic Englishman called Anthony Babington into its vortex. Twenty-four-year-old Babington had travelled extensively through Europe. And he pledged his fealty to Mary, writing to her that he and his associates intended to assassinate Elizabeth. The correspondence between him and Mary was carefully enciphered and as the smuggled letter exchanges continued (they were hidden in watertight casings in beer barrels), Mary appeared to assent, or at least to not object, to the idea of assassination. But Babington was not to know that one of his associates, Robert Poley, was a double agent, nor that the carefully smuggled encoded letters were in fact being forwarded to Francis Walsingham, Elizabeth's spy chief.

Mary used a nomenclator cipher: as well as letter substitutions, symbols were also used as substitutes for words and names. In some codes, these could be the signs of the zodiac, in others, plain numbers. Mary dictated her letters, so it was her secretary who then transformed them into cipher. The codebreakers who set to work on this ingenious and complex system, once the correspondence had been retrieved from those beer barrels, were John Sommers and Thomas Phelippes. These men occupied a crucial position in the fevered atmosphere of the time, for it was not inconceivable that an undetected plot could at some point result in the violent death of Elizabeth, and a swift Catholic reversal of the Protestant revolution. In this sense, the codebreakers held the future in their hands. As spymaster, Francis Walsingham – with his vast network of agents and double agents around the country and reaching deep into the Continent – was the prototype for a very modern kind of spy chief. He understood the vital importance of not merely deciphering codes but also deciphering the hearts that devised them.

Young Babington, and other Catholic plotters, were sentenced to gruesome deaths: to be hanged, cut down before the fatal moment and then to be cut open and have their insides pulled out before their eyes and burned on a fire. The death of Mary, Queen of Scots was also horrible: a bodged decapitation, with the axe at first connecting with the base of her skull. Hers was a life of turmoil, curtailed by codes.

20. THE ASTRONOMER AND THE ANGELS

That the Tudor court of Queen Elizabeth was in need of spies was self-evident, for hers was a throne surrounded by lethal enemies. And because the Elizabethan age still had one foot in the age of magic, sometimes those secret agents had to be versed in the hermetic ways of the occult. The philosopher, astronomer, mathematician and mage Dr John Dee – thought by some to be the inspiration for Prospero in Shakespeare's final play *The Tempest* – was one of her more mysterious agents. A scholar who travelled widely across Europe to meet with kings and emperors and to acquire strange and rare manuscripts for his own private library, Dr Dee was intensely sensitive to the uses of secret language. He would later spend time in his private alchemy laboratory trying to decipher the lost language of the angels.

So it was also natural that such a man, steeped in the darkest enigmas of the world, should also have an acute interest in cryptography. Indeed, it is thought to have been Dr Dee who acquired some of the finest encryption techniques of the time to present to Elizabeth's spymaster Sir Francis Walsingham. Dee, born in 1527, was the son of Welsh parents and educated in Essex; his father had been a courtier to Henry VIII. And Dee's fierce intellect brought him, aged fifteen, to Cambridge. Aged nineteen, he was appointed one of the original fellows of the freshly founded Trinity College. He was immersed in astronomy but also in geography. This was a time when ever more epic voyages across oceans were taking men to undiscovered lands. Dee, who would later be the first to use the term 'British Empire', had an acute interest in mapping such newly found territories. He had a profound appetite for understanding as much about the world as he could and this also meant understanding the worlds beyond, which could not be seen or known.

Dee travelled Europe extensively and lectured on mathematics in Paris. Mixed in with the purity of geometry and cartography was an underlying belief in the occult; this was not quite yet an age of pure science. Dr Dee was deeply religious and, like his fellow scholars, a practising and believing Catholic. Faith was founded upon the supernatural and so other branches of the supernatural were obvious subjects for early scientific inquiry. Language itself was an important part of all those beliefs. In cabbalistic circles, words held within themselves the mystical power to change reality. So it was understandable that Dr Dee should have been led to the practice of cryptography (in fact, the term made its debut in England in the Elizabethan age). In particular, he was drawn to the works of Johannes Trithemius, a German abbot who himself had dived deep into mysticism and language.

Trithemius had written a mighty three-volume work called *Steganographia*, which he completed in 1499 but which was not openly published until 100 years later (and even then it was swiftly banned by the Church for a further 300 years). At its heart was the communication of mysteries, and for this reason a great many assumed that it had to do with darker forms of magic. But in fact Trithemius had another aim in mind: 'I have uncovered certain ways . . . whereby I can intimate my most secret thoughts to another who knows this art, however far away I wish, securely and freely from the deceit, suspicion, or detection by anyone, using writing or openly through messengers.' He seemed to be hinting at his own knowledge of the cabala, the practice of harnessing the ancient powers locked within esoteric words. And this was a time when there was also belief in such things as the Powder of Sympathy; for instance, a weapon that had inflicted a wound was wrapped in the bloody bandage of the man that it had wounded, together with a special potion – and no matter how far away the injured man lay, the salve on the sword would start to heal his wound remotely. Could it therefore be possible to convey messages in such a curious preternatural way?

While in Antwerp in 1562, Dr Dee, himself the owner and custodian of the most awesome library in England, on the Thames at Mortlake, investigated the Trithemius manuscript and understood what it was offering. This was secret language not as an occult proposition but as a potential weapon for the state. The old German abbot had elucidated upon new principles of cryptography and new means of disguising the true meanings of important messages. Dee instantly saw the possibilities and he spent days laboriously making a perfect copy of the manuscript that he had before him. He sailed back to England and handed his precious cargo to Sir William Cecil, chief minister and adviser to Elizabeth, and the most powerful man in the realm. The codes would also have been of the most urgent interest to the secretary of state and spymaster Sir Francis Walsingham, whose career (as a practising Protestant) was founded upon ruthless and bloody suppression of any Catholic resurgence.

There were other code-setting and cracking contrivances that Dee had a passion for. One can still be found within the Royal College of Physicians today: rotating paper code discs, termed 'volvelles'. Inscribed with letters and etchings, these paper circles could be rearranged and set in a wide variety of different ways in order to form new and ever more complex codes.

None of this was a game. Life for all near Elizabeth's court was jagged with hazard and the ever-present threat of suspicion or denunciation. There was a latent violence in the state that had very much been there in the previous reign of Queen Mary, with martyrs condemned to hideous deaths by being burned alive or tortured a commonplace occurrence. Dr Dee himself had been brought before Mary's Star Chamber accused of black magic, and it had taken immense powers of persuasion to convince this implacable convocation of state grandees of the fervour of his Catholicism.

But he was to convert to Protestantism under Elizabeth and, for a time, he held great favour. He was a court astronomer and was consulted on many esoteric matters to do with the meanings of the stars in the heavens. But gradually Dee's lustre among the powerful faded and, as an old man, he became passionately interested in deciphering another form of linguistic enigma: the language that had been used by the angels. If he could find the key to this tongue, then an infinity of possible knowledge and wisdom lay before him. Dr Dee acquired a roguish associate, Edward Kelley, who claimed to be successful in contacting spirits. In Dee's alchemical laboratory, they held scrying sessions together – gazing into mirrors, or into crystals, in order to see secret spirit messages held within the depths. Dee was desperate to find and unlock what he called the 'First Language of God

Christ', 'Celestial Speech' and, more plainly, 'The Language of Angels'.

It has been suggested that Kelley had a gift for speaking in tongues and that out of this, a system or alphabet of twenty-one letters was discerned by Dee. In an effort to then break into the meaning of the words they formed, Dee inscribed the letters in code grids of forty-nine squares, 7 × 7. To pass through this gateway of understanding would give him direct access to the angels themselves, and would make him a form of prophet, transmitting their immutable and hitherto unfathomed wisdom. In other words, a process that had all the appearances of occultist ceremony – darkened chambers, candles, trances, visions in glass – was in fact a sort of alchemical codebreaking; in the end, it came down to symbols and mathematical tables.

Dee died in sad obscurity but his contribution of cryptography to the public realm via the works of Trithemius was a decisive moment in arming the English state for its empire to come. Though Dr Dee did not uncover any heavenly languages, he had foreseen that empire, and cryptography was to be one of the secret engines that would help it defeat its enemies.

21. THE DISTORTED SKULL

Ever since man took to art, paintings have held concealed meanings – even cave paintings, if a French poet called Jean-Luc Champerret was to be believed. In 1940, a band of adventurous schoolboys happened upon art that had been obscured for thousands of years upon the unexplored walls of the caverns of Lascaux, near Montignac. But this was war time and the country was over-run by the Nazis; the next people to see these extraordinary prehistoric paintings were members of the French

Resistance, on the lookout for locations that they could use as hideouts.

Among them was Champerret. According to the writer Philip Terry, Champerret made an intense study of the cave paintings, of which there are some nine hundred images in all. He considered the pieces of art long before any archaeologists could get to them. And his conclusion was that all these auroch and bison and antelope and horse images actually formed a poem. The paintings were symbols, he argued, standing in for words. When 'read' in the right way, the result was an incantation, or rite, conjuring the atmosphere of a silent prehistoric night. The idea was dismissed as preposterous in the 1940s, and yet the deeper meaning behind the configuration of all these wall images – estimated to be some seventeen thousand years old – still remains obscure.

Another artistic puzzle, this one from the comparatively recent year of 1533, hangs on the walls of the National Gallery. Although the old masters abound in all sorts of teasing enigmas, those posed by Hans Holbein in *The Ambassadors* are famously head-scratching. Here, in this sumptuous painting of two men supposed to be Georges de Selve and Jean de Dinteville, both representatives at the court of Henry VIII, is a feast of deliberate codes and encryptions. The two nobles pose against a curtain of rich green, next to a table packed with scientific instruments, globes and musical instruments; filling the foreground is the distorted image of a skull. Though the image seems so familiar, there are always new and odder details that offer themselves up when it is re-examined.

One of the men was an ambassador, the other a bishop, and the two of them were great friends. But what was Holbein suggesting with this great mass of objects around them? Why the polyhedral sundial? Why the astrolabe? What was the significance of the globes? Why, upon the lute, was one of the strings depicted as being broken? In the case of flutes, why was one missing? Why, at the very top of the left-hand corner of the painting, was there a detail of the green curtain very slightly drawn back to partially reveal a crucifix? And all this is to say nothing of the 'anamorphic' skull, that is, an image of a skull which has had its perspective distorted so that it may be viewed directly from another angle? The weirdness of the image partly lies in the intense modernity – the contrast between the straightforward expressions on the faces of the men, and the sheer intrusiveness of that illusionary skull.

Was it perhaps an allegory of heaven, earth and the inevitability of death? Did the broken lute and the pictured prayer-books hint at religious dissent (at a time when such matters were becoming ever more lethal)? The German-born Holbein was a ludicrously brilliant court painter and something of an artistic diplomat too. He had painted the portraits of the arch-enemies Thomas More and Thomas Cromwell, and he had produced a mural of Henry VIII said to be so powerful that the image alone struck quaking terror into the King's servants and courtiers (sadly, it was lost in a fire). Yet the court of Henry was a grimly dangerous place, and when Holbein was commissioned to produce a portrait of the king's next intended wife (whom he had yet to meet), Anne of Cleves, he could not have known that the fate of those around the king would pivot on the result.

Holbein was perfectly honest with the portrait. It was not especially flattering or unflattering to Anne but it did convey a sense of demure intelligence and grace. But when she and the king finally met and married, the union was an instant failure. The grossly overweight Henry, with his rotting, ulcerated leg, found *her* unalluring. The fatal end result was that his right-hand man Thomas Cromwell, who had worked so hard to secure the match, found himself in the Tower awaiting the axeman.

And a dual portrait such as *The Ambassadors* possibly had to conceal its meanings in code because, as one critic has hinted, the work was actually suggestive of an intimate relationship between Georges de Selve and Jean de Dinteville. The critic, Hagi Kenaan, wrote of the unusual circumstance of two men in the same portrait and of how de Dinteville is presented in traditional masculine fashion and pose, whereas de Selve, his hand clutching his robe, is in a more traditionally female position.

The speculation will continue as long as people gather to look at the painting: might it be as simple a matter as Holbein using all the scientific instrumentation pictured to declare his Renaissance faith in this new age of reason?

1

CAVE PAINTINGS

The prehistoric paintings discovered in the caverns of Lascaux have probably raised more questions than provided answers. What did all the shapes signify?

The shapes here do follow a structured, logical pattern. Can you describe the picture that should take the place of the question mark?

2

TUDOR SPY RING

For every question 1-10 below there are TWO quickfire clues and TWO 5-letter answers. One answer ends with the letter Y, and these are to be written in the grid, starting from the numbered space in the outer rim and working inwards.

When you have finished, a name associated with sixteenth-century espionage will be revealed reading clockwise from space number 1. However, your spy ring task is not yet over. Take the unused words, which have different endings and rearrange them, to spell out a clandestine message.

1 Tired, fatigued * Band, gathering
2 Home of monks * Comes together, encounters
3 In what place? * Fortunate
4 Forests, spinneys * Tale, narrative
5 Country, capital Rome * Traverse, intersect
6 Unpleasant * Compass's magnetic needle always points in this direction
7 Natural waterway * Tastelessly showy
8 The most famous Tudor king * Slopes which lead to natural waterways
9 Irate, annoyed * Expect, anticipate
10 Wed * In this or that place

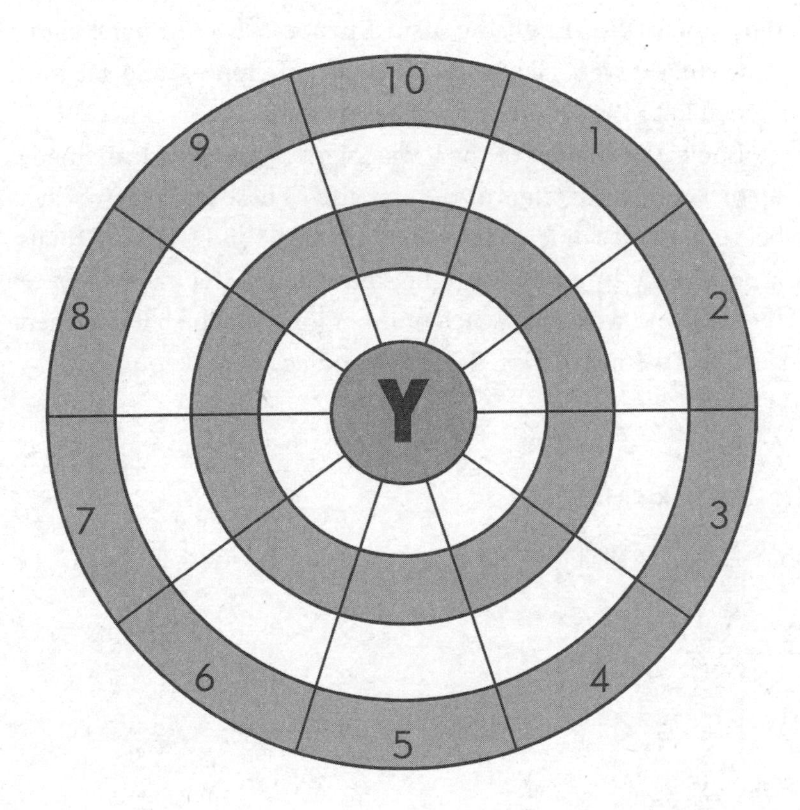
10
1
2
3
4
5
6
7
8
9
Y

3

THE QUEEN'S GAMBIT

Mary, Queen of Scots was notorious for her use of codes to plot against her English cousin Elizabeth I. Numbers, signs and symbols were all part of her world of ciphers, and she also dipped in to the world of the signs of the zodiac.

Look at the names of the zodiac signs below, which are made up of 17 different letters of the alphabet. These letters have each been allocated a different number between 1 and 17. Word totals are arrived at by adding together individual letter values. Crack the code and work out which number links to which letter, then decipher the instruction using the zodiac code as your key.

A R I E S = 40

T A U R U S = 41

G E M I N I = 55

C A N C E R = 47

L E O = 48

V I R G O = 45

L I B R A = 52

S C O R P I O = 45

S A G I T T A R I U S = 77

C A P R I C O R N = 58

A Q U A R I U S = 57

P I S C E S = 27

CODED INSTRUCTION:

6. 5. 10. 16. 8. 9.

13. 16. 1. 1. 14. 10. 16.

17. 16. 14. 11. 16.

14. 9. 15. 8. 2. 16.

4

CRYSTAL GAZING

Join Dr Dee on a scrying session as he stares into the crystals waiting to see the words that form. Fit the ten listed words into the crystals. Each word is used once and is written clockwise, a letter in each area in the hexagon shapes. Words can start in any of the spaces in the crystals. Two letters have already appeared in place.

What do the two words in the central hexagons reveal?

CALL CODE GULL HAVE JUST
KNEW PORT QUIZ SIGN TIDY

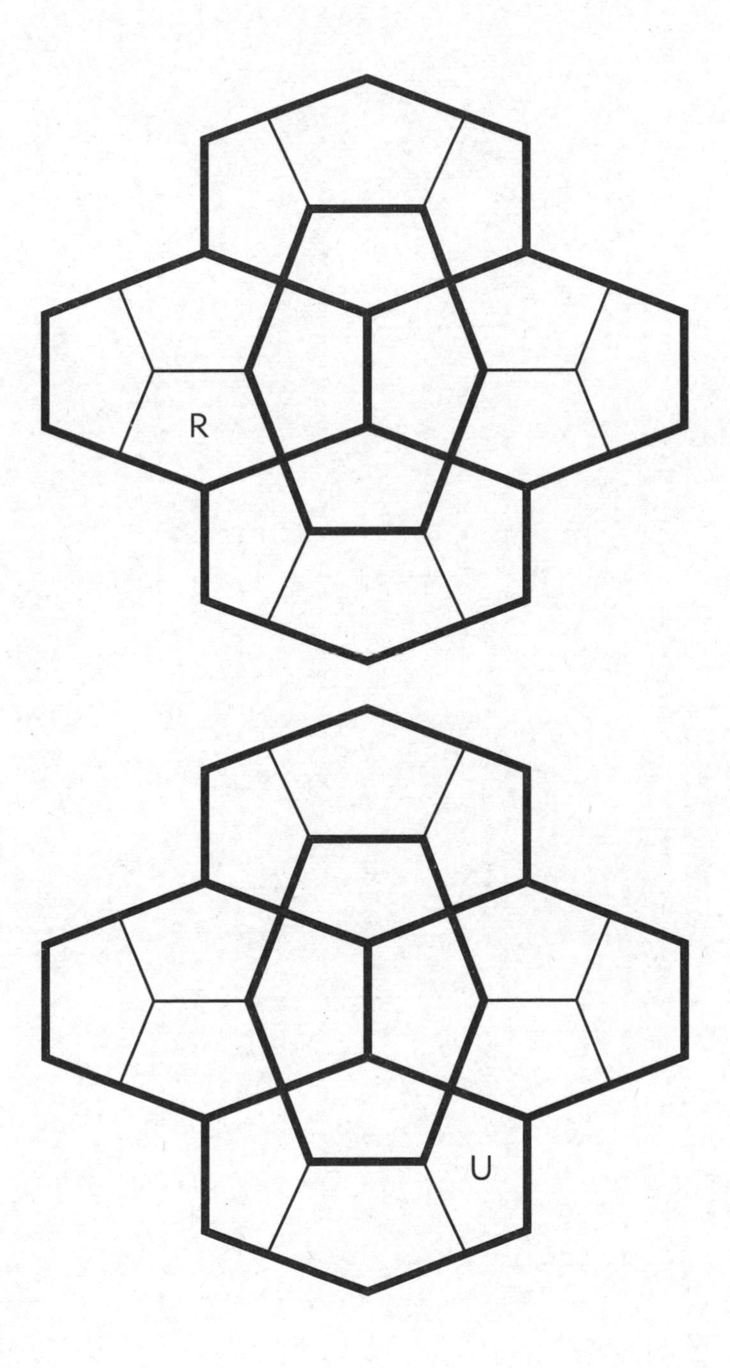
R
U

CHAPTER SEVEN

THE POETRY OF CODES

In which we explore how the language of poets has been used through the centuries to devise unusually beautiful ciphers.

22. THE VENERABLE BEDE AND THE SECRETS OF TIME

A north wind crying across the crashing sea, buffeting a stern monastery near gorsy shores, one of the few buildings in this thinly populated land made of stone. Within, against the echo of sung prayers, there is an old man in his cell, in a simple habit, writing on lambskin parchment by the light of a flickering oil lamp. He is a man who understands well the thrill of ciphers. His is a brilliant mind, set on delighting his followers with intellectual puzzles and codes that will illuminate and indeed change their world. This exceptional monk was the Venerable Bede, who lived between around 673 AD and 735 AD. To him, the entire world was a cipher to be solved and the codes he found and placed in texts were gleeful reflections of a wider quest.

When we think of Anglo Saxon England, we imagine dirty faces and wattle-and-daub houses. What we don't imagine is sparky, curious intellects or a proto-multicultural society with a delight in language and a fascination for physics as well as the new Christian religion. Bede, a deacon, exulted in riddles. It is partly thanks to him – and some head-spinning calculations to do with equinox moons – that the Easter dates that we have even now were given some anchor. There was more too – he watched the night-time sea, with the moon above its waves, and began to fathom the influence of its gravity upon the tides. He wrote compulsively and was the author of more books than Jeffrey Archer, a sum total of around sixty. The most influential of these works, *The Ecclesiastical History of the English People*, is one of the most famous and enduring histories of all time. And he also took serious delight in ciphers, as part of his huge enthusiasm for language and writing and for spreading the joy of reading.

In Anglo-Saxon society, where literacy was sparse, writing itself was a form of encryption. The letters of the Latin alphabet were, to some, as obscure and mysterious as the runic symbols used by Vikings that could still be found carved on stone (see page 18). Bede's approach to writing, as well as to natural science, poetry, music and to his own Christian faith, was about exploring and unlocking mysteries, from the currents of the rivers to the slant of sunset shadows to the age of the world itself. He wrote about time and about nature with great energy and zeal. Like so many anonymous Anglo-Saxon poets, who peppered their verses with teasing codes and encryptions that only the knowledgeable could unravel, Bede was very familiar with the cunning ruses that literate people might use to get their messages out there. Yet his philosophy was not about devising impossible riddles but about solving them. The linguistic codes he sometimes used were in the service of unlocking wider truths about the nature of life on God's earth itself.

Bede was born on Tyneside, and his religious life began early. He was inducted into the monastery of Monkwearmouth aged seven. It was a life not only of prayer and of rich chants but of the mind too. It was also a time of plague, and when he was a teenager, Bede and one other boy were left alone to sing all the devotional services when death roared through the cloisters. He was made a deacon as a very young man. It might be thought that eighth-century Tyneside was remote and wild, yet there was vigorous trade to and from mainland Europe, plus exchanges of religious and philosophical thought with everywhere from Seville to Rome. Bede was steeped in Latin and Greek and he thought and wrote naturally in several languages. Common to those who have a taste for codes and ciphers, he was also passionate about music. Adept at conveying information discreetly, as enclosed communities monasteries evolved a number of different signs and symbols. One such involved the use of hand signals. 'By this means,' he wrote, 'one can, by forming one letter at a time, transmit the words contained by those letters to another person who knows this procedure, so that he can read and understand them even at a distance. Thus one might either signify necessary information by secret intimation, or else fool the uninitiated as if by magic.'

These were hazardous times, however, and Bede, in relating the mysteries of the cosmos and of the natural earth to biblical history, ran the constant risk of being accused of heresy – a charge which for so many others would bring the most lethal consequences. Religious correspondence across the Continent sometimes contained concealed ciphers to prove the authenticity of the letter's sender. Bede was familiar with one particular secret code to keep his communications discreet, which involved taking Latin names and turning them into their Greek equivalents. Each letter of the Greek alphabet had been assigned a numerical value, and the numbers, when added together, disclosed pre-arranged keywords.

If language could contain secret hidden layers, so too could the cosmos itself. Bede's world was emphatically spherical, rather than flat, and he had the most intense interest in the dark realm of stars that surrounded it. How did these celestial spheres of light exert their influence upon the world below? And how did the light of the sun produce the rich regularity of the seasons, and the harvests they brought? To Bede, this was the cryptology of the universe where secrets of a natural world, watched over by an omnipotent deity, were ready and waiting to be gradually fathomed and decoded. Hence his involvement with what had been an impossibly complex, decades-long effort across Europe to finally settle upon which Sundays Easter would fall each successive year. Bede's preoccupation with the nature of time took him further than that. As the progress of the sun through the calendar was a mathematical phenomenon that could be calculated with certainty, so the age of the earth itself could eventually be disclosed. Bede's conclusions, including that Jesus was born some 3,500 years after the creation of the world, sparked howls of heresy and necessitated a hasty visit to the Bishop to argue with honeyed words that no harm had been intended.

Bede was a codebreaker on many levels, a man who was familiar with the Caesar Shift (see page 35) but one who also influenced the course of British and European history by suggesting all riddles might eventually be solved with the application of mathematics, logic and love. He lived in what was once termed the Dark Ages, but the light of his thought was one of the beams that led the medieval mind forward. His was a world that rang with many languages, bringing with them their own form of knowledge. And one of his central achievements, a book called *On the Reckoning of Time*, would be used in schoolrooms for centuries to come as the ultimate decoding of the zodiac and the mathematical calculation of the dates of all Christian festivals. Bede deciphered Ancient Hebrew texts and in so doing brought clarity and order to the young Christian world.

23. THE MYSTERY OF THE SHAKESPEARE CODES

'Get thee glass eyes, and like a scurvy politician, seem to see the things thou does not,' King Lear rather brutally tells the blinded Gloucester in Shakespeare's 1606 tragedy. Yet might there have been further meaning concealed within these and other lines of Shakespeare's plays? By the flickering light of oil lamps, in rooms above rowdy inns, could it have been the case that his manuscripts were being carefully written out in two subtly different forms of lettering which would indicate to spies that these exquisite lines were in fact little more than carriers of code? And more than this, as we look at the back of the man hunched over that rough wooden table and listen to the squeaky scratch of the quill (trying to ignore the beery roars resonating through the floorboards), can we even be sure that it is Shakespeare? Or is this master code-setter (and playwright) someone else entirely? Might we be looking at Francis Bacon – scholar, cryptographer and Lord Chancellor of England under the reign of James I?

The argument over the authorship of Shakespeare's plays is destined never to end (and I don't know about you, but my money is on the plain fact that actually Shakespeare *did* write his plays, and that it was not beyond the wit of a well-educated young man from Stratford to use drama to hold an amazingly clear mirror up to humanity). Typical alternative candidates – posher, naturally, for surely the author of the sonnets and the histories *had* to be posh – include the Earl of Oxford. But another frequently recurring name is that of Francis Bacon. And because Bacon was the inventor of a rather brilliant (and forward-looking) cipher, there have been conspiracy theories for years that his codes are embedded within the Shakespeare canon, that these are, in fact, the Shakespeare Codes.

With or without the excitable theories, Francis Bacon was a sufficiently substantial intellectual figure. He was born in London, close to the Thames, in 1561. And it was while he was studying at Trinity College Cambridge that this young scholar, whose family had numerous connections to the royal court, was first introduced to Queen Elizabeth, who admired his precocious razor wit. Training as a lawyer, he also travelled extensively in France, Italy and Spain, and indeed carried some diplomatic correspondence of a secret nature to be delivered to various notables. So even as a young man, he moved in a world in which cryptography was deployed and was very much an evolving art. The cipher that he was to develop, seemingly only for private use, was centuries later flourished as evidence that not only was he the true author of Shakespeare's work but was also the secret son of Queen Elizabeth and the true heir to the throne of England. The ingenious cipher was to be his way of discreetly communicating this shocking news to the world . . .

The Bacon Cipher is real enough. Rather than letter substitution, it is a means of inserting messages into any sorts of texts – even into music, if handled correctly. But it is also infernally fiddly. The first step is to arrange the alphabet into a form of binary notation, so that A is aaaaa; B is aaaab; C is aaaba; D is aaabb and so on, with all the 'a's and the 'b's used sequentially. All these 'a's and 'b's can also be expressed as binary values: A (or aaaaa) is 00000; B (or aaaab) is 00001; C (or aaaba) is 00010; D (or aaabb) is 00011.

So how could secret messages then become entwined with Shakespearean lines of soaring beauty? The trick is to use two types of font. An example: I want to convey the message 'Here's how'. And I choose to embed it in the song from *Much Ado About Nothing.* In the song, the 'a's, or '0's, will be in normal font and the 'b's or '1's will be in bold. H is aabbb or 00111. E is aabaa, R is baaaa, S is baaab, O is abbab, W is babaa. So the message would run:

'Si**gh n**o m**o**re, **S**igh no m**o**re, **m**en were **deceivers** ever . . .'

You are dividing the letters of the song into groups of five, then following the binary pattern of 'a's and 'b's. Now, of course, one objection is that the text of the song looks far from innocent and that anyone glancing at it would be struck by the weirdness of the bold/non-bold alternating fonts. But in the seventeenth century, with handwritten texts, such differences in letters could be made more subtly, with delicate pressure of the nib.

It was possibly the intriguing provenance of the cipher, which didn't appear to have been used in official circles, that piqued the interest of an American high school principal called Elizabeth Wells Gallup in the late 1800s. For some years she had made a study of the life of Bacon and was mesmerised by the bilateral cipher. Her enthusiasm was such that she began applying it to the complete works of Shakespeare (and, for good measure, the plays of Christopher Marlowe as well).

This enthusiasm gathered in intensity, and she began to discern messages and patterns hidden deep within the plays that had seemingly evaded all other eyes for over three hundred years. She became convinced that the codes revealed that Francis Bacon was in fact the illegitimate son of Queen Elizabeth I and Robert Dudley, Earl of Essex. The throne, which passed to James I, should have been his. But instead of wearing the crown, he contented himself, under the guise of 'William Shakespeare', to be the greatest playwright, the unrivalled interpreter of human nature and poet of love. Mrs Gallup was also convinced, in her readings of the seventeenth-century Shakespeare folios, that there were further plays hidden under the texts of existing plays, including one entitled *The Tragedy of Anne Boleyn*.

That manic enthusiasm, apparently, had slightly run away with her as no one else seemed able to detect the minute font changes that opened up this beguiling parallel history of England. Later, the ubiquitous American code geniuses William and Elizabeth Friedman put the Shakespeare Code to the test and also found it wanting. They could not replicate any of Mrs Gallup's findings. Yet the Bacon Cipher lives on and part of its importance, as well as being a thing of ingenuity, is that it continues to give some curious weight to the idea that Shakespeare was not the true author of his own work. The arguments will rage for as long as the world continues to watch performances of *Hamlet*.

24. THE SECRETS IN THE STARS

There persists a view that medieval London was dark, primitive and rather more brutish than our own fragrant and sophisticated era. Characterised by its maze of low dwellings, loud taverns, close lanes and dung-filled streets, we picture merchants, peasants and priests moving in crowds among animals, and smells as offensive as they come. But it is quite wrong to imagine they had no sophistication. Churches were often bright, their exteriors painted in rich colours, gardens were set out according to philosophical principles, their flowers rich in symbolism, and the herbs they grew provided natural medicine. There were visitors from around the world even then; London was and remains a great trading city. And amid the raucousness of the streets was poetry and brilliant, hilarious, heartbreaking, thrilling literature.

This was the London of Geoffrey Chaucer, and his immortal pilgrims, telling their tales en route to Canterbury. Chaucer was also something of a dab hand with cryptology. At the start of the 1400s he was as nifty with codes as any of the whizzkids who would follow centuries later. In one way, it would have been curious if Chaucer had not been versed in ciphers, for as well as being the most extraordinary poet of his day, he was also a senior courtier to Richard II, a diplomat, and by extension most likely an occasional spy. While it was *The Canterbury Tales* that guaranteed him cultural immortality, he authored other fascinating works, one of which is thought to have been an astronomical study: *The Equatorie of the Planetis*.

A document that described the workings of a star-observing instrument called an equatorium, this work was, in essence, a means that might be used to track the movements of the heavens. It has recently been ascribed to others, including a monk, yet it bears a reference to Chaucer himself and seems related in some ways to a similar work of his called *A Treatise on the Astrolabe*. At the heart of the *Equatorie*, accompanying some mathematical tables, is an enciphered section: an extraordinary fragment of medieval London code.

There are six ciphers, in which the letters have been substituted with symbols. One of the ciphers when decrypted was revealed to mean: 'This table servith for to entre into the table of equacion of the mone (moon) on either side.' In other words, there was nothing here of concealed politics or statecraft. Yet at the same time, the author – whether Chaucer or the claimant monk – will have understood that even astronomical knowledge could be sensitive at times. To map the movement of the sun and the moon and the stars, and to give an empirically observed account of the heavens, in a religious age, was touching the frontiers of belief and faith. Astronomy and astrology went together in medieval times and the belief that the spheres in the skies above could light the destinies of those below seemed natural and pervasive.

The use of a cipher might have been the equivalent of a masonic nod and a wink, a signal to other readers of similar intellect and interests that this text was explicitly for their eyes. Codes are in part about secret relationships like these, conspiratorial and conversing furtively. All that said, mathematics, algebra and geometry would also have been encipherment enough should a number of characters from *The Canterbury Tales* have encountered them.

Chaucer belonged to a time where English as we know it in its modern form was just emerging into being, in a city that had resounded with Old English, French and Latin for centuries. Chaucer was not only the first poet to write in recognisable English, he (or his monkish rival) was also the first to take this new language and transmute it into magical encoded symbols.

25. THE LIFE THAT THEY HAD

Their mission in the Second World War was to set Europe ablaze. Their methods involved infiltration and sabotage. Using disguised explosives, they destroyed everything from railway lines to factory buildings, steel and concrete consumed in molten blasts. And at every moment, waking and sleeping, they were being hunted down by the Nazis. The secret agents of the Special Operations Executive were parachuted into Europe and frequently were all alone in hostile territory, on missions that they were very aware could end with their deaths. What made these agents, women and men alike, even more remarkable was that in peacetime, no one would ever have imagined that this latent fire lay within them. And as befitted such unusually courageous recruits, their system of codes involved both lateral thinking and an element of genuine romanticism.

Perhaps the most famous of all SOE's wartime recruits was a young woman called Violette Szabo. She had been born in Paris in 1921 to a French mother and a British father. By the age of eleven, she and her family had moved to London. Violette was very active: a skilled ice-skater, a long-distance cyclist (as so many were in the 1930s) and even in her teens she had a good eye for shooting. Yet the jobs she got after leaving school gave no clue to the trajectory of her future. She worked at Woolworth's in the centre of London, for a corset-maker in South Kensington and for a department store in the then drab suburb of Brixton.

When war came, Violette threw herself into the effort, first becoming a land girl and then working at an armaments factory in west London. It was at a special Bastille Day parade in London following the fall of France that she met Etienne Szabo, and theirs was a romance of intense speed. They married quickly, and before long he was dispatched to North Africa to join the fight against Rommel's forces. Violette gave birth to their daughter, Tania, in 1942 and not long after she learned that her husband had been killed in action.

There was no possibility that she would accept passive widowhood. She was determined that she should do everything in her power to fight the enemy. And so it was, possibly because of her fluency in French, that she was speedily and silently recruited for the Special Operations Executive, leaving her baby daughter behind with relatives. Training in the Highlands of Scotland was gruelling and she would have been put through intense all-night survival sessions on the moonlit moors, with tutorials on cutting throats and shooting people dead. She and her fellow recruits were being sent into France both to help local Resistance groups and to initiate their own spectacular acts of sabotage against the Germans.

Throughout her training, Violette was inducted into the discipline of cryptography, an especially tricky and fraught skill when out in the field, or in hiding, with hostile soldiers in close pursuit. One of the great geniuses of the SOE was Leo Marks, who supervised all sorts of ingenious means by which secret messages might be smuggled (including squares of silk). He also oversaw a clever cipher operation which meant that codes used by secret agents were drawn from books – anything from popular novels to soliloquies from Shakespeare. The book codes operated on the principle that agents would first be using the same edition of a book as a reference, so that the page numbering and the paragraphs were all the same. And it was from these that keys were derived. For instance, the first few letters of the second paragraph of chapter three of *Bleak House* could be used as the starting point for a transposition cipher.

There was one slight difficulty, as captured agents found to their horror; the Germans assigned to cryptography were every bit as literate and cultured as their British equivalents. This meant that the basis of codes could sometimes simply be guessed (especially if the agents concerned had selected Hamlet's 'To be or not to be' soliloquy as the key). This meant that for codes to be truly effective, yet also be memorable enough that the agents did not have to write them down (with all the dangers of discovery that brought), they had to be original. And so it was that within the Baker Street headquarters of the Special Operations Executive a special team was formed to devise new poems that could be learned off by heart.

One of the more famous examples worked by means of sheer vulgarity. It was doggerel poetry that ran: 'Is De Gaulle's prick / Twelve inches thick? / Can it rise / To the size / Of a proud flag-pole? / And does the sun shine / From his arse-hole?' Again, from this, the key to a cipher could be agreed beforehand and worked out quickly in frequently frightening conditions out in the field. Yet the poem that Leo Marks gave to Violette Szabo as her code key had a breathtaking romanticism. It ran: 'The life that I have / Is all that I have / And the life that I have is yours. / The love that I have / Of the life that I have / Is yours and yours and yours. / A sleep I shall have / A rest I shall have / Yet death will be but a pause / For the peace of my years / In the long green grass / Will be yours and yours and yours.'

As with the other poem codes, it was possible to devise a quick transposition cipher using agreed letters or phrases as keys, and because the poem was original, no enemy codebreaker would be able to find any reference to it. Yet this particular code poem acquired its own terrible haunting depth. Violette had completed one extraordinarily brave and successful mission in France under an assumed name (she had even managed to get to Paris), and the time came for a second parachute drop deep into the heart of the enemy's operations. On this occasion, she worked with the local Resistance on plans to harry and undermine the German military in the wake of D-Day. But this time around – possibly thanks to an ankle injury made worse by the parachute landing – she got caught up in a gun battle in a field and was taken prisoner.

The ordeal that she and other SOE agents went through was dreadful. They were relentlessly tortured by the SS and forced to undertake hard labour and endure starvation in Ravensbrück concentration camp. Then, just a matter of weeks before the Allies were able to crush the Nazis, Violette and the other agents faced cold execution in the camp yard. A gun was aimed at the back of her head and fired.

Though she faced a terrible end, the story of Szabo's courage and self-sacrifice was not lost and held some of the hope that the poem ends with. Her little daughter was taken to Buckingham Palace for a ceremony in which her late mother was awarded a posthumous George Cross. And that poem code, 'The life that I have', became inextricably intertwined with her name. When her story came to be dramatised in a film a few years later, Leo Marks gave permission for the poem to be used, as long as it was attributed to Szabo's late husband, Étienne. Thus this code acquired an extra layer of romanticism.

And the poem codes also came to symbolise a certain nobility about cryptography, and the cause that it could serve. Here were codes that were being used by heroes, in the most extreme jeopardy, poems that could make the difference between life and death.

1
VERSE AND WORSE

My first is in CHARM
And also in HEART.
My second is in STOPS
But it isn't in START.
My third is in MEET
But it isn't in LEAVE.
My fourth is in CARING
But it isn't in GRIEVE.
My fifth is in STRANGER
And also in FRIEND.
My sixth is in SCENES
But isn't in SEND.
My seventh is in HUMBLE,
And also in MEEK.
With the clues from my verse
Do you know what I seek?

2

POETRY QUEST

Words linked to poetry are hidden in the box of letters. All words are in straight lines and can go horizontally, vertically and diagonally. They may read forwards or backwards. However, there is one word in the list which does not appear in the grid. Which word is it?

ALLITERATION
CLASSICAL
COUPLET
ELEGY
EPIC
HAIKU
IAMB
IDYLL
LIMERICK
LINES
LYRIC
METRE
ODE
ONOMATOPOEIA
PASTORAL
POEM
QUOTATION
REFRAIN
RHYME
SONG
SONNET
STANZA
TITLES
TRIPLET
WORD

O	I	U	K	I	A	H	E	P	I	C	N
C	N	M	G	Z	J	T	E	A	Z	O	O
L	S	O	N	N	E	T	K	S	I	D	I
A	S	A	M	L	O	C	R	T	R	U	T
S	T	E	P	A	I	S	A	O	G	L	A
S	A	U	N	R	T	R	W	R	M	T	T
I	O	D	E	I	E	O	H	A	E	N	O
C	J	M	T	T	L	Y	P	L	T	L	U
A	I	L	I	A	M	B	P	O	R	Y	Q
L	E	L	Z	E	N	I	E	S	E	R	V
S	L	H	O	P	R	E	F	R	A	I	N
A	D	P	R	T	Y	G	E	L	E	C	A

3

MOVING WORDS

Here are some moving words in which the letters themselves have been moved. The lines have been written in rows reading across. There are no spaces between words. The columns of letters 1 to 6 have moved in order, and then the rows have all changed position. You may be moved to distraction trying to solve this.

Have a go at working out the lines of poetry from the letters.

There are a couple of clues at the foot of the page if you need them.

R	P	T	N	A	O
R	A	N	R	B	O
O	D	L	A	S	L
E	G	C	S	A	A
A	M	O	I	N	S
W	E	O	S	N	T
I	R	N	K	O	E

CLUE 1

The final letter is an E.

CLUE 2

The first letter is an S.

4

BRUSH UP YOUR SHAKESPEARE

Here are three different sets of Shakespeare quotations, each set needing a different code to decipher successfully. You might not get the answers in 'one fell swoop' (*Macbeth*), but when you have finished, you will conclude that '*All's Well That Ends Well*'!

CODE A

1 T I W F O L U O S E H T S I Y T I V E R B

2 W O R R O S T E E W S H C U S S I G N I T R A P

3 E G A T S A S D L R O W E H T L L A

CODE B

1 P O D F N P S F V O U P U I F C S F B D I E F B S G S J F O E T

2 U I F X J O U F S P G P V S E J T D P O U F O U

3 P C S B W F O F X X P S M E

CODE C

1 I S F H M A U K S E I S C P B E E A T R H E E S F H O A O K D E O S F P L E O A V R E E P S L H A A Y K O E N S

2 T S H H E A C K O E U S R P S E E A O R F E T S R H U A E K L E O S V P E E N A E R V E E S R H D A I K D E R S U P N E S A M R O E O S T H H A

3 I S A H M A A K M E A S N P M E O A R R E E S S I H N A N K E E D S A P G E A A I R N E S S T H T A H K A E N S S P I E N A N R I E N S G

5

STAR LINES

Like Bede and Chaucer, many of us are fascinated by the night sky and the arrangement of stars. Arrange the numbers from 1 to 12 so that the total number in every line is the same.

Some numbers are already in place.

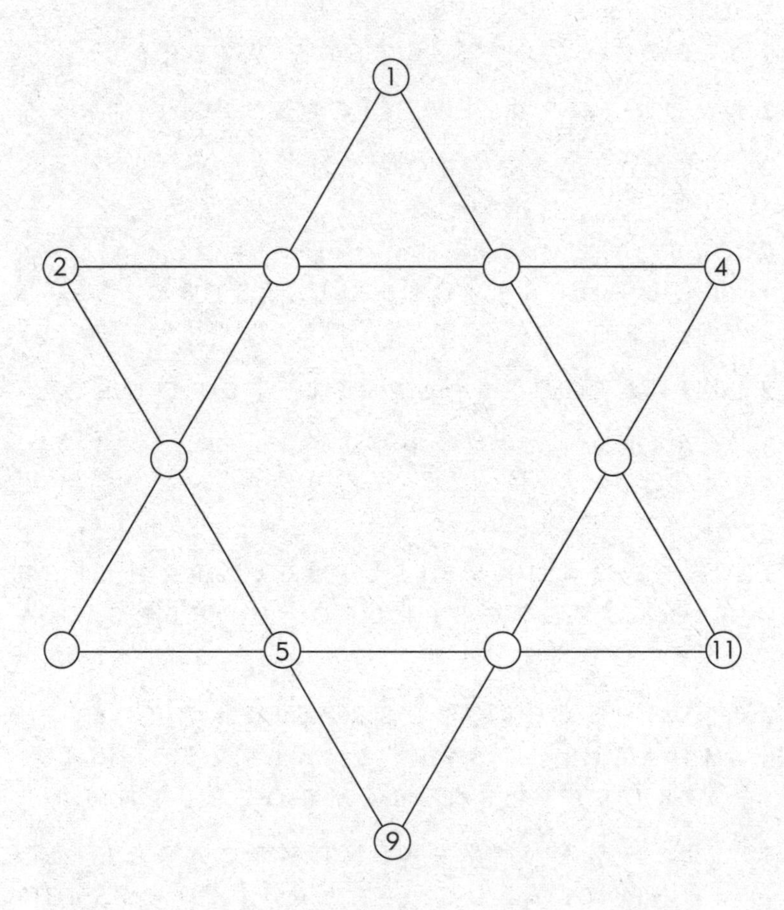

6

VALENTINE

From the eighteenth century onwards, the sending of Valentine cards became more widespread. Puzzles and codes were often employed so that secret messages could be exchanged. The sender of this card was well advised to conceal the message!

The message can be deciphered by moving from letter to letter in any direction except diagonally. Each letter is used once and once only. Where to start? That's for you to decide.

T	E	D	E	R	E	R
S	L	V	S	E	S	A
A	O	I	O	E	M	E
R	M	A	R	V	D	I
E	U	Y	L	O	O	N
B	O	Y	Y	E	V	T
L	U	E	O	U	O	L

CHAPTER EIGHT

VICTORIAN ENCRYPTIONS

In which the history of Victorian pomp and majesty is explained in its amazing zeal for inventing and cracking ever more fiendish codes.

26. THE IMPERIAL ENIGMA

Amid the breathtaking mountains and the fairy-tale ruins, the land was punctuated with the corpses of British and Russian soldiers. The peninsula was Crimea. Over one hundred and fifty years ago, the war that was fought there between Russia on one side and the Ottoman Empire, France and Britain on the other was a storm of mighty cannons and horseback sabre charges. Of course, there were also codes; back in London, a very singular genius, in the shape of a man who invented the principle of the computer a century too soon was making his own contribution to the Crimean War through cryptography. For Charles Babbage, the whole business of codes and decryption was irresistible – he had been delving into ciphers ever since his schooldays.

In so many ways, Charles Babbage was a man ahead of his time. Born in 1791 and having discovered a rapacious appetite for mathematics in his boyhood, he was something of a dreamer. Indeed, in his early career, he was too much of a free spirit to ever settle into any proper full-time job or academic position. The idea of 'scientist' as a professional calling had not quite solidified by this period. Even though his talent was recognised in universities from Cambridge to Edinburgh, and at the Royal Society, it took the young man some time to find his focus. When he did, it was intense. By the age of thirty-seven, he was Lucasian Professor of Mathematics at Cambridge, the founder of the Royal Astronomic Society, a theorist on the new landscape of factories and how to make them work efficiently, a member of The Ghost Club, dedicated to the supernatural, an enthusiast for diving bells, altimeters and study of the new means of creating multicoloured lights on stage, and also a visionary who was imagining an extraordinary machine that might carry out a range of complicated calculations.

The Difference Engine was Georgian technology; Babbage started to build it in 1822. The original aim was astronomical mathematical calculations. A small working model of it suggested a beautiful mass of brass and iron, with interlocking cogs and numbered wheels, powered by hand-cranking. But despite a large government grant, there were difficulties in creating a full-scale operational machine, and after many years, the enthusiasm abated (only to be sparked again by the genius of Ada, Countess of Lovelace, among others). But the knotty logical problems posed by assembling a Difference Engine (and the subsequent Analytical Engine) were in a way related to Babbage's enthusiasm for cryptography. He inhabited a world of intellectual conundrums that would surely yield to the power and force of human logic.

'Deciphering is, in my opinion, one of the most fascinating of arts, and I fear I have wasted upon it more time than it deserves,' he wrote in *Passages from the Life of a Philosopher*. 'I practised it in its simplest form when I was at school. The bigger boys made ciphers, but if I got hold of a few words, I usually found out the key. The consequence of this ingenuity was occasionally painful: the owners of the detected ciphers sometimes thrashed me, though the fault really lay in their own stupidity.' He also wrote: 'There is a kind of maxim amongst the craft of decipherers . . . that every cipher can be deciphered.'

Babbage shared this enthusiasm with a friend called Charles Wheatstone, and together they turned codebreaking into a kind of game or intellectual hobby. Texts would be encrypted as a challenge and Babbage would unravel them with the ease of a cat attacking a ball of wool. 'I am myself inclined to think that deciphering is an affair of time, ingenuity, and patience,' he wrote, though he added, 'and that very few ciphers are worth the trouble of unravelling them.'

But was his talent noted in the higher reaches of the British establishment and the military? In 1853, there came the first booms of war from the region of the Black Sea. A dispute between the Russian and the Ottoman Empire would lead to Britain and France coming to the aid of the Ottomans and getting pulled into an extraordinarily complex and bloody conflict that came to be known as the Crimean War. As well as the tangle of great empires throwing their soldiers into the lines of cannon fire, this was noted as being the first war in which modern technology began to play a serious role, including that of telegraphy.

The cipher most widely used at this time in military and diplomatic circles was the Vigenère Square (see page 45 for a profile of its illustrious inventor). This polyalphabetic cipher, deceptively simple in principle, had defeated an array of the finest minds for decades. It had been declared utterly unbreakable.

There was no logical chink that could be exploited that would help to prise it open. Yet it has been speculated that Charles Babbage had mastered it prior to the conflict in Crimea. The theory links him to Admiral Francis Beaufort (progenitor of the Beaufort Scale, measuring wind force, as well as a distinguished naval captain, hydrographer and the conduit for Charles Darwin joining the voyage of the *Beagle*). By the time of the Crimean War, Sir Francis was long-retired and very elderly, but he too was something of a cryptographer. Did he and Charles Babbage excitedly discuss the means by which the Vigenère Square could be mastered? Sir Francis had engineered a reverse version of this cipher, which he used in personal correspondence. The theory goes that Babbage – possibly on the recommendation of Sir Francis – was inducted into secret cryptography work for the British military, and that he had long succeeded in breaking apart Vigenère Square codes by means of fresh mathematical analysis. Thanks to him, it was now possible to read Russian military ciphers.

The very fact that the codes had been cracked would have had to be kept super-secret, for if the Russians, or any other power that employed the same cipher methods, knew that a hyper-enthusiastic British mathematician had broken into them like a burglar, then they would immediately change all their systems and all advantage would be lost. Yet there was a clue that slipped out in 1854, when another keen amateur cryptographer called John Thwaites decided to challenge Babbage to a codebreaking duel. Thwaites also used the Vigenère Square, confident that his opponent would have to concede defeat. Instead, to his astonishment and irritation, Babbage decoded his message with some flair. This then suggested to the wider community of mathematicians and talented amateur codebreakers that Babbage had seized this Holy Grail of cryptography. And it does now seem inconceivable that someone of Babbage's talent could have been overlooked for a conflict as serious as the Crimean War.

As a postscript, the secret of Vigenère did get out a few years later, but the method was published by one Friedrich Wilhelm Kasiski. The cat was out of the bag, but still across Europe and Eurasia this cipher continued to be used for some time. As for the Crimean War itself, by 1856 profound weariness had set in on all sides and peace was as complex as the factors that had led to the entanglement. But the British occupation of the port of Sevastopol, and the ruination of Russia's navy, exposed fissures in Tsarist Russia which would widen as the decades passed.

Babbage would have had little idea of the immortality of his legacy in both the realm of cryptography and the future world of computing. He never got round to publishing a mooted philosophy of codebreaking. But he did write of its human factor: 'One of the most singular characteristics of the art of deciphering is the strong conviction possessed by every person, even moderately acquainted with it, that he is able to construct a cipher which nobody else can decipher. I have also observed that the cleverer the person, the more intimate is his conviction. In my earliest study of the subject I shared in this belief, and maintained it for many years . . .'

27. PLAY UP, PLAY UP, AND PLAY THE GAME!

The Victorian era saw a flowering in the widespread use of codes both in war and in romantic love. What had once been the hidden preserve of embassies and spies gradually became a subculture in its own right. It was throughout this period that the British Empire, stretching across continents, was becoming ever more established. However, conquest and dominion was continually threatened by rebellion and rival powers. Those who were subjected to the will of British overlords frequently found that other powers were seeking to overthrow them and take their place. In India and Afghanistan, the continual rivalry between Britain and Russia came to be termed 'The Great Game', and secret codes and ciphers played a strong part in espionage among those rich, mountainous landscapes.

But there was also a domestic element to the delight in cryptography among a great many people back in Britain. Newspapers and journals were proliferating, and letter-writing was so fundamental to personal communication that households received three deliveries a day. The coded messages that could be sneaked into personal columns, or sent via the specific angles of postage stamps on envelopes (as we shall see below), were in curious ways related to the encryption games being played by spies in the Hindu Kush, as later immortalised by the novelist Rudyard Kipling. In short, this was an age in which secret codes became not only integral to military strategy but also, in a new age of mass communication, a clever means for young couples to send each other disguised declarations of love (and lust). The Victorians took cryptography and made it a mainstream pursuit.

Although the image of empire has changed rather radically in the last 100 years, its history continues to exert the most powerful fascination. In fact, when it came to one brilliantly innovative cipher development, British diplomats initially shied away from it on the grounds that it was *too* ingenious, and therefore too complex to use. It was the invention, in 1854, of a brand-new encryption system by Charles Wheatstone, an acquaintance of Charles Babbage. The hook was, rather than encrypting single letters, it would encrypt pairs of letters, cleverly arranged on a grid. This meant that the encoded messages would not contain frequently recurring single letters that could prove the loose thread to unravel the entire text.

Wheatstone was a formidable eccentric, who hailed from Gloucestershire. His incredibly active intellect roamed across a variety of disciplines, from music to electricity generation. He was rather reclusive, but the time alone brought forth a range of wonders, including a very early form of 'piped' music, 'the enchanted lyre', which would transmit performances from one venue to another. In 1834, he devised an experiment involving wires, sparks of electricity and mirrors, all of which was to measure the speed of electricity through wires.

His enthusiasm and inventiveness also brought him into the quest to develop electrical telegraph systems. And on top of all this, he devised an encryption method that would remain in use for almost a hundred years, not only throughout the British Empire but subsequently in two world wars. Strikingly, though, this code came to be known not by *his* name but by the florid surname of an MP who tried to persuade his sceptical Foreign Office bosses that this system really was both brilliant and practical.

That floridly named MP was Lord Playfair, who was also Postmaster General, and a scientist too (who was apparently keen to begin use of chemical weapons during the Crimean War). He and Wheatstone were friends, and Lyon Playfair was tremendously enthused by Wheatstone's demonstration of his new encryption system. It was made up of a very simple-looking 5 × 5 grid (which allowed for twenty-five letters, meaning that I and J could be interchangeable). By encrypting pairs of letters, called bigrams, and marking out their positions within the grids in rectangles, you could introduce levels of deep complexity to foil the efforts of decoders. It was a system that could be used with pencil and paper, out on battlefields or mountain missions.

Explaining it to senior officials was quite another matter though. Lord Playfair gathered together some ministers and officials and took them through it, complete with a demonstration of the 5 × 5 grid. The ministers were not keen. But, Lord Playfair exclaimed, he could prove how easy it was to learn the system by trying it out on schoolboys. Three out of four, he promised, would have mastered it within fifteen minutes. That might have been so, retorted the Foreign Office secretaries, but 'you could never teach it to attachés'. Their reluctance was misguided, and gradually the system was taken up in the latter half of the nineteenth century. However, instead of being known as the Wheatstone Code, it was called the Playfair Cipher, simply because he had pushed so hard for it.

Indeed, the Playfair Cipher was to prove remarkably durable. From the heights of the Himalayas to the sultry soil of southern Africa, the system enjoyed a historical moment before the advent of machinery that might be able to pick through its layers and combinations with rather more speed and patience than harassed subalterns. Even in the age of Enigma, at the time of the Second World War, Charles Wheatstone's clever grid was still being used out in the field by various agents. Indeed, Bletchley recruits such as Captain Jerry Roberts had to spend some of their time mastering the mathematics and logic of a system that had been invented 100 years earlier. By that stage, it was Double Playfair because the Germans had found a way of enhancing the complexity by introducing a second square of letters. And so it was that Captain Roberts found himself prising open messages sent by Nazis using a method that had only a few years previously featured in a Dorothy L. Sayers mystery novel, *Have His Carcase*. As was frequently the case, the line between popular fiction and real codebreaking was very thin.

Back in that high-Victorian summer of empire (some territories of which the British had seemed to hold on to almost accidentally), one hugely gifted novelist conjured a vision of espionage amid timelessly old landscapes. Rudyard Kipling's *Kim* was the story of an orphaned teenage boy in India working for the British military, while setting out on a more resonant quest with a wise lama who seeks the 'River of the Arrow'. As well as being one of the first spy novels and both a gripping adventure and a swooning evocation of an extraordinary land and its people, it was a work of some profundity about identity and time. Here was where the idea of 'The Great Game' was first popularised – a mighty geopolitical struggle between the Russians and the British for control of Asia. This was a realm of constant skirmishes and tensions, and of secret documents and messages being carried through valleys and across snowy mountain ranges. Characters such as Mahbub Ali were also known by their codenames (in his case 'c.2.1B'). Kipling himself had a fascination for codes, later devising his own puzzles aimed at children (including a drawing of a pictogram code on an elephant's tusk).

But the uses of codes by the Victorians were not confined to the military. They had their romantic purposes too. There was 'the language of stamps'. The angles at which postage stamps were affixed to envelopes each had their own discreet meaning. A tilt slightly to the right meant 'Forget me not'; a tilt to the left 'Come soon'; upside down meant 'Do you remember me?' and upside down and tilted to the left meant 'With all my heart'. Thus were the unspoken flirtatious thoughts of apparently repressed Victorians expressed, with humour and a hint of daring. More sharp tilts decoded meant 'Do you love me?' and 'Answer at once', which gave the game of coded love a sense of extra urgency. Yet at the same time, the General Post

Office knew of these codes too, so any young lady or gentleman receiving a letter with a skewed stamp would have received very knowing looks from their postman.

28. FAR FROM ELEMENTARY, MY DEAR WATSON!

"It is certainly a rather curious production," said Holmes. "At first sight it would appear to be some childish prank. It consists of a number of absurd little figures dancing across the paper upon which they are drawn. Why should you attribute any importance to so grotesque an object?"

"I never should, Mr Holmes. But my wife does. It is frightening her to death. She says nothing, but I can see terror in her eyes."

Thus begins 'The Adventure of the Dancing Men', a pleasingly intense tale in the volume *The Return of Sherlock Holmes*, published in 1903. That the author Sir Arthur Conan Doyle should have his hero wrestling with a macabre encryption problem was not in itself remarkable, but the texture of the story, and the ingenious nature of the code itself, proved in its own way resonant and influential.

A Norfolk landowner and his American wife are in receipt of bits of paper which have 'dancing' stick men drawn across them. That the wife clearly understands their meaning indicates that, conceivably, this is a code that can be broken by others. But the story (spoiler alert!) swiftly darkens further when the husband is found shot dead in a locked room and it appears that the room's only other occupant, his wife, must be the guilty party. But Holmes, who has begun to see a way into unravelling the code of the dancing men, perceives that there is much much more to the sinister affair than that. And he sets out to reveal the real killer by adopting the code himself, using it to send a message, via the dancing men, that will act as a trap . . .

Secret codes in popular fiction, despite their mass entertainment value, were curiously influential on real cryptographers. Bletchley Park's Dilly Knox had a youthful fascination with Holmes's intellectual methods (see page 197 – telegram section) and took issue with another of the code puzzles in the oeuvre – the dried orange pips sent to victims-to-be in 'The Five Orange Pips' – convinced that he had spotted a logical flaw. Equally, Women's Royal Naval Service volunteers for codebreaking in the war, such as Ruth Bourne, had devoured spy stories involving complex encryptions. In the case of Ms Bourne, when she arrived at the hyper-secret Bletchley Park she found herself resolutely unsurprised by the nature of the work that was revealed to her.

Cryptography and the works of Sir Arthur Conan Doyle had also formed part of the literary and imaginary landscape of a senior figure in Naval Intelligence who in 1952 would create a spy whose very serial number was itself a form of cipher. Ian Fleming had been brought up on spy and detective thrillers, from Holmes to the (proto-fascistic) Bulldog Drummond, and understood the world could be a murky place filled with secret signs and ciphers. His immortal creation, James Bond, 007, was a government assassin rather than a cryptographer, but there were times when his adventures revolved around codes. In the 1957 novel *From Russia with Love* Bond is lured into a Soviet trap baited with an encryption machine called Spektor (not a million miles from Enigma, at that post-war stage still very much a secret). Then the 1964 novel *You Only Live Twice*, set in Japan, features another encryption system, MAGIC 44 (a much cheekier lift, since MAGIC was indeed one of the classified terms used in the US for the Japanese code system).

As with most of his fiction, Fleming drew heavily on real

life, even for Bond's more baroque adventures. And there was the wartime acknowledgement that every secret service had to have firewalls of unbreakable codes. That in turn had been fed by his youthful reading of material such as Arthur Conan Doyle. For what makes 'The Adventure of the Dancing Men' so compelling is the sense that Holmes is in a race against time to fathom this creepy code of gesticulating, dancing stick men. Death is fast approaching and Holmes's success or otherwise might determine who lives and dies.

And the method by which Holmes shatters the problem and turns the tables on the villains? The same principle of frequency analysis that had been hit upon amid the golden sands of the Middle East in the ninth century. That some of the gesturing stick men in each message are identical suggests that they stand for the most commonly used letters. When the story was originally published, each twist of the tale was accompanied by fresh illustrations of the dancing men together with the terms that had been deciphered.

There were other ciphers to be found in the adventures of Holmes (warning: spoilers lie ahead, so if you are new to Holmes's literary exploits, you might want to glide over the next few paragraphs). In 'The Adventure of the *Gloria Scott*', another Norfolk-dwelling unfortunate, the elderly Mr Trevor, is in receipt of a menacing text that to everyone else looks innocuous. He is in the company of a curiously ill-mannered, lazy and drunken butler called Hudson. The note comes in the form of a letter that reads: 'The supply of game for London is going steadily up. Head-keeper Hudson, we believe, has now been told to receive all orders for fly-paper and for preservation of your hen pheasant's life.'

At first the code, which brilliantly does not look like a code, refuses to yield its inner meaning. But Holmes at last unlocks it and finds that it is a blackmail letter, referring to an incident far in the past involving prisoners on board a ship. The method? The key is that one should read every third word, starting with the first. In this way, the text goes: 'The game is up. Hudson has told all. Fly for your life.'

Then there is the cipher that opens the novel *The Valley of Fear*, which at first sight is an impossibly complex proposition. Holmes has an inside man within Moriarty's criminal organisation who is going by the name Porlock, and Porlock has sent him a letter, which runs thus:

> 534 C2 13 127 36 31 4 17 21 41
> DOUGLAS 109 293 5 37 BIRLSTONE 26
> BIRLSTONE 9 47 171

On this occasion, Holmes tells Watson that 'there are many ciphers which I would read as easily as I do the apocrypha of the agony column; such crude devices amuse the intelligence without fatiguing it'. This code, however, is a different matter. Anticipating the love poetry codes of the Second World War's Special Operations Executive (which took their keys from real poems of the seventeenth century), this too 'is clearly a reference to the words in a page of some book' (see page 148).

But which book? Porlock's second letter written from within Moriarty's HQ tells Holmes that for fear of discovery he dare not send on the key. Is the code, therefore, wholly unbreakable? It requires one specific book as its key, but can it be deduced? It must be a book that Holmes owns, otherwise how would the cipher ever work? And the starting point is the number 534. The working hypothesis is that this is a page number, so it must be a very big book. And the following C2? Watson suggests 'chapter two'. Holmes retorts that for chapter two to come that late in any volume would be testing. But then Watson gets it: Column

2! So, it must be a large book printed with two columns on each page. That really does narrow it down. The Bible? Unlikely to be seen anywhere within Moriarty's HQ. An almanac then? What about *Whitaker's Almanack*?

A moment of inspiration unlocks the most apparently intractable cipher. The other numbers refer to the position of the words on the page. 'There is danger – may come – very soon – one Douglas – rich – country – now – at Birlstone House – Birlstone – Confidence is pressing.' The result, as Holmes tells Watson, is not exact, but the meaning of it is plain enough to him. This 'Douglas' who lives in the style of a 'rich, country' gentleman at Birlstone is to be targeted in some fashion.

In Conan Doyle's fictional world, the entire landscape of city and country was a cipher on a grander scale. Ordinary suburban villas, small country houses, lanes and busy streets on the surface may have appeared perfectly normal but were actually, in their ways, encryptions as well. Beneath that surface lay other hidden meanings; the keys were the clues which Holmes's gimlet eye explored every corner for. And in a sense, this attitude went on to inform the approach of real-life codebreakers. Theirs was an ability to not only analyse seemingly random lines of letters and numbers but to tilt the head and look at the world that produced these symbols from quite a different angle.

29. DICKENS AND THE DEVIL'S HANDWRITING

As well as conjuring a carnival of immortal characters, Charles Dickens lived his own life to a peak of colourful intensity. He worked at a speed that others could not comprehend. This velocity not only applied to his outpouring of fiction but also extended to editing popular journals and embarking on intensive speaking tours. There was something almost feverish about his compulsion to write swiftly. And this was manifested most visibly in his use of a form of code that has only recently been deciphered.

Furthermore, this code later evolved, and by the twentieth century was forming the cornerstone of business, medical and journalistic affairs. It was, of course, shorthand, but the term could cover a wide variety of bewildering encryptions. The Charles Dickens code was among the most tantalising of them all. He himself described his own shorthand as 'the devil's handwriting' and the fact that it firmly resisted decoding by others – until a team led by the University of Leicester finally prised open its secrets – says much.

The surviving example that the international team of literary experts worked upon was known as the 'Tavistock letter', which was sent by Dickens on headed notepaper bearing his address in Tavistock Square, London. The coded writing was made in blue ink and the angular squiggles at first glance look too unruly and diverse to belong to any form of ordered system. But they very much were, and a clue to their provenance can be found in his semi-autobiographical novel *David Copperfield*. In it, the young hero, launching himself upon the tides of the world, becomes a trainee journalist and as such must learn shorthand in order to accurately transcribe matters such as parliamentary proceedings. This was true of young Dickens too. He worked in an age before voices could be recorded and it was imperative that word-perfect transcriptions were committed to paper, often with fast-blunting pencils. The effort on the writer's wrist was intense (as was the subsequent business of deciphering the shorthand), but as a discipline, it sharpened focus to an extraordinary degree.

Shorthand – abbreviated writing using symbols to stand for

combinations of letters or words – was scarcely new. A form of it had existed as long ago as the Roman era, as scribes struggled to capture senatorial speeches and instructions. In medieval England, new systems evolved. By the seventeenth century, the noted diarist Samuel Pepys was using one developed by Thomas Shelton, not merely for speed but also because it was a handy way to keep private notebooks encrypted. There is no doubt that prying eyes would, on the whole, be baffled by the coded writing.

But the most famous form of shorthand – a form of cipher that came to be used all around the world – was invented by a Wiltshire man in 1837, the year that Queen Victoria came to the throne. He was an English teacher called Isaac Pitman (who would later be knighted for his achievements). Pitman was quite a pioneer in a number of ways. He was a dedicated vegetarian and he was passionate about spelling reform. Like Dickens, he too desired to work at speed; 'time saved is life gained' was apparently his motto. His invention was a form of phonetic shorthand which he termed 'sound-hand'. Based in the spa town of Bath, where he taught in a small school, Pitman started up an innovative postal course by which means he taught his new method of shorthand. It was based upon the way words sounded – dots, light dashes, circles and curves representing various vowel and consonant combinations. And though it was complex, the fact that it could be studied via postal courses (rather than face-to-face tuition) gave it a widespread appeal that, over the course of decades, would become global. Even before this worldwide success, the business had attracted so many customers that Pitman gave up school teaching to devote himself to it. He had a brother in Cincinatti, and so it was that Pitman shorthand crossed the Atlantic as well.

If anyone can be said to have perfected shorthand, it was

Pitman. The method spread so far into the professions that the very name acquired its own distinct fame. Doctors used it for medical notes, secretaries for taking dictated letters from bosses, journalists for reporting and for interviews. Until recently, it was still a firm requirement for journalists, even after the advent of mini tape-recorders. There was always the chance that the recorder could break down, and what would they do then? In addition to this, if newspapers were taken to court, the courts would require that shorthand notes were turned over for examination. For a younger generation armed with smartphones, the very idea of Pitman may seem antediluvian. However, it genuinely had a claim to being a code that made the world go round. Speed and efficiency were combined (and Marxists might say that, in this sense, it was also a very capitalist cipher, maximising efficiency and thereby profit).

Before all this, in the early nineteeth century, when young Charles Dickens was a court reporter making notes on innumerable trials for the newspapers, there was a system called brachygraphy, which he described as being very difficult. This did not stop him adapting it in his own fashion, making it even more difficult. The Dickens Code, as presented in the mysterious letter, was very much the author's own individual invention, its Byzantine complexities some distance beyond the Pitman shorthand.

The letter has been pored over for decades. There was even a prize of £300 offered for decoding it. Composed in 1859, it was written when Dickens was part way through composition of *A Tale of Two Cities*. And it was thanks in part to a Californian computer technical support specialist called Shane Baggs that its secrets recently began to be stripped away. Although not a Dickens devotee, Mr Baggs was very keen on any cipher challenges and this historical example appeared irresistible. He spent six months of his after-work hours subjecting the letter to analysis until finally some cracks began to show (enabling him to win the prize).

The subject of the letter was a complaint. Dickens had tried to place an advertisement in a newspaper for a new literary publication, but for some reason the clerk on duty had rejected it. Dickens was writing to higher authorities to express his grave displeasure and also to ensure that the advertisement ran in a future edition. Even with all of this disclosed, however, it seems that there are still fragments of the code in this letter that continue to resist analysis.

30. THE TSAR'S VISIONARY

Through the crunching snow that lay uniformly across beautiful cities from Warsaw to Riga, St Petersburg to Moscow, there moved a figure who attained near-legendary status within the feared secret police of the Russian tsar. He travelled widely across steppes, his expertise sought after in many corners of what was then the Russian Empire. His capacity to descend into mystical trances of thought, unable to be reached by anyone, gave him a slightly supernatural aura (some years before the advent of the monk Rasputin into the tsar's court). His skill was that of codebreaking. Ivan Zybin had an uncanny ability to see deep into the heart of any cipher and wrest the truth from its deceitful coils. At a time of constant threatened revolution, terrorism and severe oppression, the secrets that men encrypted posed direct and regular threats to the powers that held Russia in such a tight grip. Zybin could not stem the tides of history, but he was destined to become part of the country's secret past.

Like their counterparts in western Europe, the Russians in

the nineteenth century had been adept at establishing 'black chambers'. Secret departments were set up where diplomatic letters and other protected correspondence could be sent, opened, copied, resealed and resent, while cryptographers worked on the ciphers they had harvested. By the time of the reign of Tsar Alexander II, such codebreaking efforts were taking on swarms of underground revolutionary movements. Having acceded to the Russian throne in 1855, Alexander had at last bowed to the modern world with the emancipation of the serfs in 1861, and the effect upon Russia's rural society was significant. At last, peasants were free to own their own property and to marry whom they wished without seeking the permission of patriarchal and oppressive landowners. But for huge numbers of younger radicals, this and other reforms could only be viewed as the first steps towards freedom from all aristocratic oppression.

The tsar became the focus of a range of assassination attempts, some where the putative killers confronted him with a pistol in public spaces – such as the busy 1867 World Fair – and others where they attempted to blow up buildings that he was in (such as the ballroom of the Winter Palace). In the end, death came in St Petersburg. The tsar was in his coach, accompanied by several Cossacks, when a young man stepped forward from the crowd with a package tied up in white cloth. He threw it under the horses' hooves – and it exploded.

But the carriage had been specially modified to make it, to an extent, bulletproof, and though one of the Cossacks inside had been killed, the tsar was able to emerge from the wreckage. At this point, though, a second young assassin from the same revolutionary group, The Peoples' Will, materialised. And with the uncanny cry, 'It is too early to thank God!' he threw another bomb. This time the tsar was caught fully in the blast and those racing to his aid were horrified to see the blood pouring from the stumps of his shattered legs. With his death came the end of any discussion of reform. Instead, his heir Alexander III, and thereafter Nicholas II, brought in a new era of terror and oppression, encapsulated by their use of the sinister secret police, the Okhrana.

As the twentieth century dawned, the codebreaking arm of the Okhrana had become a world leader in a technique that others might profitably have copied had they been able to afford it: cheating. Its agents simply went to work on the staff of various embassies and offered them vast sums of money (bribes) to buy their codebooks. There would always be some employee somewhere who would be happy to accept a monstrous bribe for an act of espionage that could not necessarily be traced back. In this way, various western diplomats became wary of sending certain enciphered telegrams. The assumption was that the Okhrana would be ready to pounce. And yet it was not always so efficient, and the Russians' own use of codes was not tremendously sophisticated. Ciphers were dissolving before the scrutiny of all sides.

Standing apart from all of this was a genuine genius. According to author David Kahn, Ivan Zybin was 'fortyish, tall, thin, swarthy, with long hair separated by a parting and with a lively and piercing look'. Kahn also cited the admiring memories of Okhrana chief Pavel Zavarzine, who had the opportunity to watch this wayward eccentric genius at work. One cipher in 1911, the work of a revolutionary group, was composed mainly of fractions, and was wholly befuddling to the decrypters in Moscow. Zybin was summoned and as he set to work, his commandeered desk began to fill with paper upon which was scrawled great masses of formulations. The spy chief offered the great codebreaker dinner and had to offer again and again before the wholly absorbed genius would even look up. He had to be forced away from his desk and sat down at the dinner table, whereupon he was offered a bowl of broth. He drank it quickly before turning the empty bowl over and continuing his work on its surface, with a pencil. In a way, it is easy to see the parallels between this immersive eccentricity and that displayed a little later in Britain by Alfred Dillwyn Knox (see page 199).

The point of the broth episode was that it was at the spy chief's table that Zybin had his epiphany and unlocked the cipher. The solution was part mathematical and part to do with a substitution code, and the message that it spelled out concerned a shipment of explosives in innocent-looking boxes that were being delivered to Kyiv prior to the tsar's state visit. Having accomplished this intellectual feat, according to Kahn, Zybin then sat back down at the dinner table and happily and calmly ate the remaining courses.

Yet this weird genius was of little use to the tsar or his war effort as the world hurtled into conflict in 1914; Zybin's responsibility was decrypting, but it turned out that Russian military codes were many degrees weaker than those of their enemies. There were points at which entire battles had been decided simply because the Russian forces had either used easily solved codes or, indeed, altogether failed to encrypt messages at all. In the vortex of the Great War, defeat had a thousand causes. Though in terms of codes that changed history, the tsarist efforts went part of the way towards doing so precisely because they did not work. Their opponents were decrypting them so fast that they were sometimes in receipt of intelligence before the commanders that the messages had actually been intended for.

Then came Vladimir Lenin, who was inserted into 1917 Russia like a plague carrier, his desired revolution soon to bring an empire to its knees, then to be slowly, agonisingly reborn as the Soviet Union. With the bloody Bolshevik seizure of power in November 1917, the old tsarist world was turned inside out. Nevertheless, there were some elements of continuity even in this violent disorder. One of those was the central team of cryptology experts that had been formed across the years, many with aristocratic lineage who had been asked to serve in the Black Chamber because of their dazzling educations and easy proficiency with a number of European languages (something the poor serfs could never have hoped to acquire).

But this was an era of radical modernisation and the Soviet Union went to work on expanding and improving the nature of its own cipher systems, as well as continuing to work to demolish those of others. This was a new world of equally new alliances and rivalries. It has been suggested that the old guard of Black Chamber cryptographers were deemed to be necessary even into the 1930s – the talent for deciphering being so rare as to require constant nurturing – but the darkness of Stalin's paranoia crept into the rooms where these experts worked, as it did throughout the Soviet Union. The 1930s and the 'purges' were a time of mass slaughter under his totalitarian rule, and no one was safe from the weird, atavistic 'purifying' show trials and subsequent executions. It has been further suggested that among the many victims of Stalin in 1937 were the older codebreakers – among them, the once-venerated, quirky cipher-breaker Ivan Zybin.

1

THE GHOST CLUB

The Ghost Club had its beginnings at Trinity College Cambridge in the 1850s, where all things ghostly were discussed and debated. In London, its members included Charles Dickens, Sir Arthur Conan Doyle and Charles Babbage.

Ghostly goings-on suggest phenomena which cannot be seen, yet witnesses feel there is something there. Look at the message below. You can't see the message clearly but there is something apparent. Can you crack the code and reveal the message?

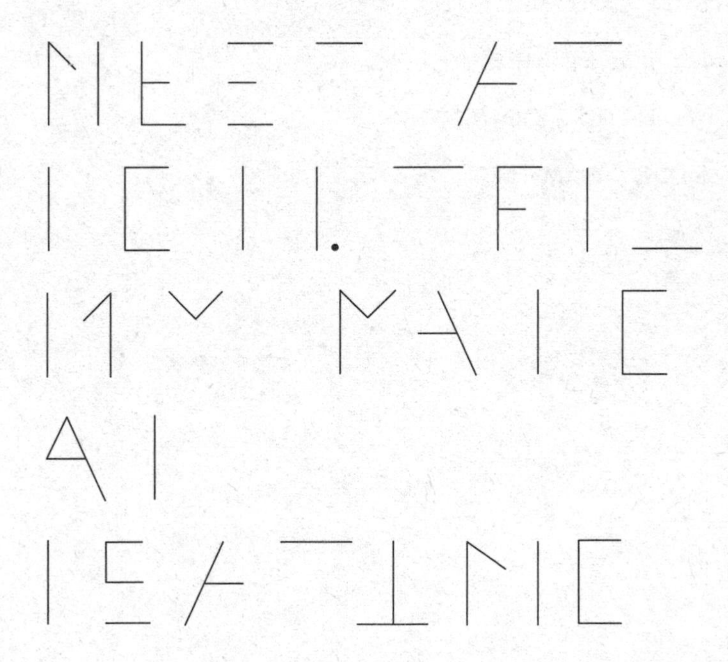

2

AUTHOR, AUTHOR!

The Victorian era brought with it a fashion for making anagrams from the names of the rich and famous.

In this puzzle, some eminent nineteenth-century authors have had their names turned into anagrams. Can you find the literary celebrities of the period? Each solution is made up of two words.

1. ANYONE COLD?
2. SICK LARCHES END
3. MY LONE TRIBE
4. WILL LIKES COIN
5. STORY HAM HAD

3

SCIENTIFIC CIPHER

Scientific advancement in the Victorian era marched on at a pace. The Victorians were fascinated by its progress. In 1869, at the very heart of the Victorian period, a Russian chemist by the name of Dmitri Mendeleev created what become known as the Periodic Table. What is the message concealed here in a very scientific cipher?

PHOSPHOROUS, IODINE, CARBON, POTASSIUM,

URANIUM, PHOSPHOROUS,

THORIUM, IODINE, SULPHUR,

NITROGEN, OXYGEN, TELLURIUM,

BORON, YTTRIUM,

NITROGEN, OXYGEN, OXYGEN, NITROGEN.

4

ANAGRAMMATIC

The Victorians didn't have crosswords as we know them today, but they loved anagrams. In this crossword, each solution is an anagram of its clue, i.e., the solution uses the same letters as the clue but the letters are in a different order. Beware! Some clues can give several different solutions but there is only one way they can all fit into the grid.

ACROSS

7 GREASE

8 OLD MAN

10 MELISSA

11 U BOAT

12 SEAT

13 EDICT

17 CHAIN

18 VOTE

22 ALERT

23 DANGERS

24 RANTED

25 PARTED

DOWN

1 BEAR CAT

2 EMPIRES

3 EVENS

4 DECLARE

5 RONDO

6 SITED

9 PAGE ONE IS

14 GARNISH

15 REVEALS

16 ESCORTS

19 TABLE

20 DUSTY

21 GENRE

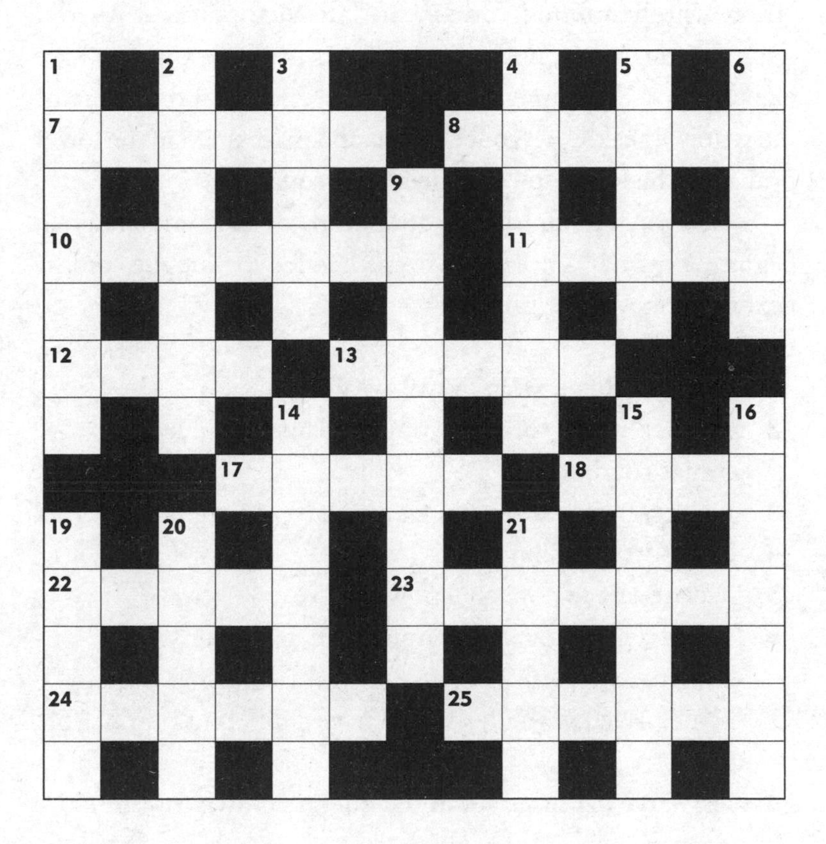

5

CODED QUOTE

In this puzzle, the code is part of a quotation by a poet and writer who first found fame in the late Victorian era. Answer the questions horizontally in the upper grid. All answers have eight letters. Column A reading down will reveal the name of the writer. Take the keycoded letters and place them in the lower grid, e.g., the first letter you need is in square C6.

When you have finished, a quotation will be completed that begins, 'A man…' Three vital items needed for supplies on the next mission will be revealed!

1 Outer garment worn in wet weather.
2 Another item to protect your head and shoulders in similar conditions.
3 In a descending direction from the top of an incline.
4 The sport of sailing.
5 Declare, make publicly known.
6 Part of a traditional telephone, which contains the earpiece.
7 Game played with tiles which have a different number of dots.
8 Very large diamond set in the queen's crown for the 1937 coronation of George VI.
9 Not fully grown or developed.
10 Dispensary, chemist's shop.
11 Serenade, choral outpouring of affection (4.4).
12 Living for ever.
13 Writer of fiction.
14 Official in charge of a prison, a ruler.

	A	B	C	D	E	F	G	H
1								
2								
3								
4								
5								
6								
7								
8								
9								
10								
11								
12								
13								
14								

C6	C10	E8		D3	H9	C11	D13	D10		E3	G12	C14	E2	
H1	D5	F7		B9	E5	G10	C8		H6	D11	A3			
C3	F13	F14	G6		F12	F1	B7		C9	H2	G11	H10		
C2	G14	B13	A8	E11		D12	H8		E4	B14	B11			
B12	F9	G5	D4		G1	C7	C12	A2	A13	E6	H13	C1	B3	G4

6

DECIPHERING DICKENS

In Charles Dickens' *Pickwick Papers*, Miss Gwynn was the writing and ciphering governess at Westgate House School, Bury St Edmunds. Carry on her good work by solving this cipher.

The letters in Miss Gwynn's name have been replaced by symbols. The symbols remain constant throughout. If MISS GWYNN is written as:

💣 ✋ 💧 💧   ☝ 🕈 ✡ ☠ ☠

who are the other Dickensian characters?

1 ☠ ☜ 🕈 💣 ✌ ☠   ☠ ⚐ ☝ ☝ 💧

2 ☠ ✌ ☠ 👍 ✡

3 ☞ ✌ ☝ ✋ ☠

4 💣 ✌ ☼ ☹ ☜ ✡

5 ❄ ✋ ☠ ✡   ❄ ✋ 💣

6 🏱 ☜ ☝ ☝ ⚐ ❄ ❄ ✡

7 💣 ☼   💣 ✋ 👍 ✌ 🕈 👌 ☜ ☼

8 💧 ✡ 👎 ☠ ☜ ✡   👍 ✌ ☼ ❄ ⚐ ☠

7

THE DANCING MEN

In the 1903 story 'The Adventure of the Dancing Men', Sherlock Holmes noticed that certain illustrations of the figures were repeated. He surmised that this was a substitution code with a specific drawing representing a specific letter. Imagine if the great detective and his trusty colleague, Doctor Watson, were involved in another case of dancing...

The pair visited a rural village in Kent where a group of Morris dancers are performing outside the local inn.

'Don't see why these chaps do this,' murmured Watson.

Holmes held his hand up for silence. There were five dancing men. In turn, each went into the centre and the other four danced round him. The man in the middle stood still, holding his two ornate handkerchiefs in his hands. This circuit was repeated, so that each man took two turns in the centre before the end of the dance.

'How very interesting,' remarked Holmes, as he politely applauded the performers. 'Did you deduce the message, Watson?'

'Message? All I saw was some fellows holding handkerchieves!'

'Exactly! The dance gave out a coded message. Each of the dancers in the middle was using semaphore. The positioning of the handkerchieves represented a letter of the alphabet.'

'Ingenious, Holmes, but I cannot recall all the shapes that make up the semaphore code,' returned Watson.

The detective carried on his explanation, warming to the theme. 'All that you need to know is that seven different letters of the alphabet were used. The letter used first was repeated as the fifth letter. The sixth letter was repeated as the ninth letter, and the seventh letter was repeated as the eighth letter. No doubt you observed that the final letter, an E, was the only example of a letter taken from the first half of the alphabet, and that the first five letters shown were in fact the last five in alphabetical order, with no letter after the single U being displayed.'

'Come on, Holmes!' cried the exasperated Doctor. 'You can tell me the message.'

Holmes paused and smiled at his colleague.

'Can I?' he asked.

8
THE PLAYFAIR CIPHER

It was said of this nineteenth-century system that schoolchildren could pick up the hang of it within 15 minutes. But it does have subtleties which take a little while to get used to. This is in some ways a cross between a Polybius Square and the Vigenerè Cipher, and this system had the advantage of being both eminently portable and also – if you didn't have a clue about the keyword – fiendishly complex (if not impossible) to unscramble. One of the reasons is the element of bigrams, which will be explained shortly.

The keyword is central here. The 5x5 grid is – like the Polybius Square – there to be filled with all the letters of the alphabet, with 'i' and 'j' sharing a square. The difference here is that instead of just filling each letter in sequence, the code keyword comes first. Say, as in the first puzzle on page 195, the keyword is 'nordic'. First inscribe these letters so the sequence left to right, row by row, would read 'n', 'o', 'r' 'd', 'i/j', and then in the second row 'c'. These in place, you now fill the rest of the grid with the remaining letters of the alphabet, so the second row would continue with 'a' 'b', 'e' and 'f' (no letter should be repeated). The bottom row will read 'v', 'w', 'x', 'y', 'z'.

The brilliance of Playfair lies in its elegant twist: letters are paired. You will see that the first encoded message starts 'wd', 'si', 'an', 'dp' and so on. Each pair is inseparable. Take 'w' and 'd'. If you look for them on your grid, you will see that 'w' is second from left on the bottom row, and 'd' second from right on the top row. Draw an imaginary rectangle that encompasses these two letters. What you are now going to do is look for the letters that mirror them within that rectangle. In the case of 'w', that mirror is second from the right on the bottom row: 'y'; and in the case of 'd', that letter is second from the left on the top row: 'o'. The next pair is 'si'. 's' is in the middle of the fourth row; 'i' at the end of the first row. Again, draw the imaginary rectangle that encompasses them both – and look for their mirror letters in that rectangle. Thus: the mirror of 's' is 'u' and the mirror of 'i' is 'r'.

And there you have it. You have found your first word: 'Your'.

Not all the letter-pairs form rectangles. Some lie on the same row. If they are on the same horizontal row, the technique is to take the letter immediately to the left. If vertical, take the letter immediately above. In the case of repeated letters – i.e. 'pp' in grid 2 – select the diagonal mirror letter in the grid. Note: all the puzzle messages may end with a final 'x' to signify a stop.

And again, when devising your own codes, simply reverse the above procedures (and when letters are on the same lines, reverse the direction, so that you take the letter immediately to the right, or below).

We have devised five puzzles, each of which will need their own 5x5 square. Either solve the puzzles with a pencil, and then erase the letters in the squares or take a ruler to draw fresh squares. This will also help if you are devising your own Playfair code. The triumph of Playfair was that without the key, the enemy would have next to zero chance of even beginning to penetrate it. There was eventually a mathematical formulation that could detect the frequency of letters, but that would offer little comfort in the field. This is why it was still in use throughout the Second World War, and why Bletchley experts were trained in techniques to drill into it. Brilliant Home Secretary-to-be Roy Jenkins cut his Bletchley codebreaking teeth on Playfair. There was Double Playfair too, adding extra layers of complexity.

But even the system as it stands is a genuine feat of Victorian ingenuity.

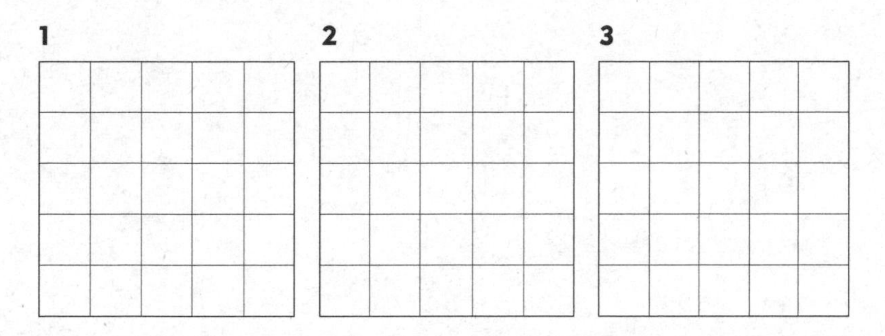

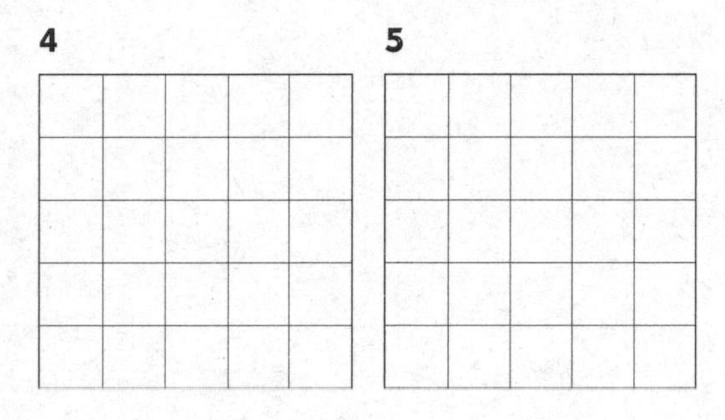

1 WD SI AN DP BA UD UR OR TK RW
KEYWORD: NORDIC

2 ME LM RA DN UD HK PP YH DU
KEYWORD: LOCOMOTIVE

3 NU EB ZU ML QK DO ML DC GT IV
KEYWORD: REMBRANDT

4 IF DV HM HF AR BM TM PB BW
KEYWORD: MONA LISA

5 NC BE YB EG BC PS NI XA
KEYWORD: DYNAMITE

CHAPTER NINE

CIPHERS OF THE GREAT WAR

In which, amid all the terrors and the bloodshed of the First World War, codebreakers sometimes enjoy triumphs that prove historic turning points.

31. THE TELEGRAM THAT CHANGED EVERYTHING

The blood had run in red streams into the wet mud. The bodies of thousands lay whole or in fragments in dead landscapes. Men were living in trenches and tunnels, desperately trying to hold on to their sanity. By January 1917, the war that had swept the world had already destroyed too many lives to count. In the same month, there was an incredible development which would be instrumental in turning the course of the conflict. It involved a brilliant feat of codebreaking by a man who was sometimes affectionately known as 'Dormouse'. This cryptographic feat had repercussions that would be felt far beyond the war and so, in a sense, it helped change the entire political shape of the world. And all because a German politician had no idea that telegrams sent across the Atlantic were routed through a small-scale station in a tiny seaside village in Cornwall.

In Europe, the Great War had begun in 1914 with great patriotic parades, men and women lining the streets, full of optimism and giddy excitement. No one could have envisaged the filthy reality of the Battle of Verdun, or the Somme. America had stayed neutral. By 1917, there were those in Britain and France who desperately wanted this to change. Meanwhile, in Germany, as High Command planned a U-boat assault on Atlantic shipping to hit British supplies, the German State Secretary for Foreign Affairs, Arthur Zimmermann, was cultivating his own stratagems, one of which was encouraging an alliance between Germany and Mexico. The idea was that, should the US enter the war, Germany would join forces with Mexico to help it win back its former lands in Texas, Arizona and New Mexico.

On 17 January 1917, Zimmermann sent a coded telegram to Heinrich von Eckardt, the German ambassador in Mexico, outlining this idea and instructing him to speak to the president. This telegram was routed via the German embassy in Washington DC. Sending encryptions via diplomatic cables through neutral countries seemed a guarantee of security. Days later, to Zimmermann's astonishment and horror, his telegram was being waved around like soiled laundry before the world. Before long, in part thanks to the outrage of the American public, the US entered the war. And by doing so, it then went on to determine the shape of Europe after it had won helping sow the seeds for future conflicts.

But how was Zimmermann's telegram captured and deciphered? It was the work of a secret Whitehall department – the forerunners to Bletchley Park – referred to as 'Room 40'. A hotbed of English eccentricity, it was run under the gimlet eye of Admiral William 'Blinker' Hall, director of the naval Intelligence Division (his nickname was an un-politically correct reference to his pronounced facial twitch, likened by some to naval signal lamps). It was a department of pronounced personalities. One codebreaker there, Alfred 'Dilly' Knox, compared himself and his colleagues to figures in *Alice in Wonderland* and used the riddles of Lewis Carroll as a template for lateral cryptographic thinking. Then there was 'Dormouse', a brilliant man called Nigel de Grey, who enjoyed art and amateur dramatics. These two men understood precisely the world-altering potential of the German communication that had been laid before them.

Telegrams across oceans were sent using underwater cables and neither Arthur Zimmermann nor indeed the Americans had realised that the British had found a way to tap these before the cables even entered the water. There was a relay point at Porthcurno in Cornwall, a charming village stretching around a yellow sandy bay with cliffs. It was a sort of wild, storm-blown staging post for international telegrams that would be zipped over to New York and Washington. Tapping the wire at Porthcurno had yielded Zimmermann's letter and Nigel de Grey, using the template of previous German diplomatic messages that he had decoded, patiently set to work together with Dilly Knox. The results, as they began to emerge from that labyrinth of encryption, were so explosive that Room 40 and the government were initially not quite sure what to do with them.

'We propose to begin on the 1st of February unrestricted submarine warfare,' was one of the sentences that emerged. 'In doing so, we will endeavour to keep America neutral . . . if we should not succeed, we propose an alliance on the following basis . . .' This was a plot against America, in black and white.

'We had got a skeleton version,' de Grey later recalled, 'sweating with excitement because neither of us doubted the importance of what we had in our hands . . . As soon as I felt sufficiently secure in our version, even with all its gaps . . I ran all the way to (Blinker's) room . . . I burst out breathlessly, "Do you want to bring America into the war, sir?"' Blinker very much did, as did the British government. The fresh infusion of new supplies and the weight of manpower could (and did, in the end) prove decisive. The immediate difficulty was in getting the intelligence out to the world without the Germans realising that their codes had been broken (which would lead to them ramping up security on future messages, thus depriving Britain of intelligence). The operation was smoothly thought through. British intelligence would contrive a means to make it look as if the contents of the telegram had leaked out in Mexico. The next step was to find a way of controlling the wave of outrage that would come.

US president Woodrow Wilson had just the year before been re-elected on a pledge that America would stay out of the war. On 17 February 1917, UK intelligence ensured that the decrypted telegram was passed to him. News of it also made its way into the US newspapers. Thanks to the intensified submarine war, there had already been a slow burn of hostility to the German military, but the telegram fanned it into a blossoming flame. By April, President Wilson was standing before Congress calling for war to be declared, on the grounds that 'the world must be made safe for democracy'. This was a turning point that echoed throughout the coming century. It was America emerging as a superpower, willing to police the world.

Naturally, Nigel de Grey's pivotal role in this most extraordinary codebreaking coup was kept deep within the shadows (as indeed was Room 40 itself). But it was to have something of an impact upon his own life, even in the inter-war years, as he went off to head up an art concern called The Medici Society.

He was still very much on the codebreaking radar and when, in 1938, the distant thundering booms of approaching war could be heard once more on the wind, he was back within the fold. The Dormouse was drawn into the regenerated Room 40, now the Government Code and Cypher School, relocated some forty miles outside of London to a smart country house and estate called Bletchley Park. There he remained in the directorate throughout the war, helping to oversee the department's conversion into the world's most formidable codebreaking factory.

And he was a strong presence after the war too, helping to remould the department once more as the world now faced the possibility of Cold War nuclear conflict. De Grey was instrumental in setting up what we now all know as GCHQ. So in this sense, this charming man – who loved appearing on the stage, and who would paint special birthday cards for each of his grandchildren – really could claim to have been one of the most influential codebreakers of all time.

32. WITCHCRAFT IN THE TRENCHES

In amid the squalor and horror and exhaustion of one of the most terrible conflicts the world had seen – treeless horizons, blasted vistas of mud and bone – there were, startlingly, outbreaks of inventiveness. In the spring of 1918, the German war machine, like all others, should have been almost paralysed by the nihilism of what it was doing, and yet it was in this unlikely environment. on the Western Front that a brand-new cipher was conceived. It was the brainwave of a twenty-six-year-old signals officer called Fritz Nebel. And it was a new kind of code designed for ease of use with wireless telegraphy while on the move or close to battle. It came to be termed the ADFGVX Cipher and there was a time when it was believed to be wholly uncrackable.

More cumbersome and less successful cipher systems were already in use in this filthiest of wars. The trench codes relied on field telephones, carrier pigeons and a variety of other means that could be intercepted with some ease. And all sides were using a truly bewildering number of different codebooks in rotation; officers had to know which days to use which particular ciphers. In the ghastly circumstances of the trenches, these codebooks too were vulnerable as any push by the enemy that seized and held precious yards of muddy territory could see the codebooks fall into the hands of the opposition.

Nebel's idea had a certain swish elegance, employing as it did a device from a much earlier century: the Polybius Square, that 5 × 5 grid into which all letters of the alphabet were placed (I and J shared a square). This was adapted to 6 × 6, which would also give room for the numerals 0–9, as well as making it even more fiendishly complex. Along the top horizontal line and left-hand vertical line of this 6 × 6 grid were written the letters ADFGVX. To begin encrypting a letter or a number, you simply read off its co-ordinates on the square.

The reason for using the letters ADFGVX? In Morse code, they were individually very different in the way they sounded. This meant that there would be less chance of signals officers making mistakes when they were transmitted. The codes were translated into Morse and sent in five-letter groups. And there was some confidence that they were simply too complex for any mind to crack. However, one such mind did exist within the ranks of the opposing French army. And what unfurled during the grim carnage of 1918 was a curious form of mental duel, between the man who had invented the code and the man who would drive himself to the very limits of his sanity to unravel it.

Captain Georges Painvin was a thirty-two-year-old cryptologist attached to France's Cabinet Noir (Black Chamber) throughout the course of the Great War. In his formative years he had been both an enthusiastic mathematician and an accomplished musician (he won a prize for his cello recitals). As a young man, in the early years of the twentieth century, he was in and out of military service and when he was not gaining artillery experience, he was pursuing his academic career, eventually graduating as an engineer and becoming a professor of palaeontology. Painvin's uncanny skill with ciphers had been noted not long after the outbreak of the Great War. It was he who took on the apparently impenetrable Imperial German Navy ciphers, and those of the Austro-Hungarian Navy too. As we shall see a little later, his cryptanalytic path crossed that of the exotic dancer and spy Mata Hari (see page 205) and he was in part responsible for her fate. But one of the central points about Painvin was that many of those who watched him at work could not begin to fathom how he was conjuring meaning from fractal chaos. There seemed to be something a little supernatural about his ability to think his way into a cipher, without any sort of external aid.

The sheer ingenuity of the ADFGVX cipher was the greatest challenge he faced, though, and there are those who have suggested that the breaking of it in the late spring of 1918 thwarted a terrible German offensive against Paris. There was, rather poetically, an intercept station for coded messages based at the top of the then relatively new Eiffel Tower. It was via this post that the crucial ADFGVX ciphers began coming through, and the opposing side knew that to unlock them would unlock the secrets of German positions and intentions.

Painvin had deduced that the cipher system involved a Polybius Square, and that it was arranged upon a 6 × 6 grid.

Even knowing all this was not enough for most people to even dream of finding a way into it, but the knowledge that he was working against the most terrible countdown, and that attack could come at any time, meant that Painvin threw himself almost into a fugue state of concentration for a day and a night, with no rest or let-up. After twenty-six hours of what one colleague described as 'witchcraft' (mathematics and probability computations, carried out mentally), Painvin had succeeded in reconstructing the grid and the key to it. And from this came forth a message written in clear German. It ran (in translation) as follows: 'Speed up supply of ammunition. If not seen also during the day.' The Eiffel Tower listening station had the location from which the message was sent and now the French forces knew where to send their defence.

But this mighty, focused effort – the climax of his cryptanalytical war – almost broke Painvin and he had a form of collapse or breakdown. He was insensible as, just a few months later, the horror of the war was brought to an end in November 1918. He remained exhausted and unable to take on any tasks for some months after that. It was as though the efforts on ADFGVX, as well as all the other German ciphers, had somehow used him up or depleted him. Recovery was slow but he eventually returned to industry in the peacetime world. (Regrettably, twenty years later, there was some ambiguity about his role in France in the Second World War, with some suggestion that his industrial expertise was lent rather too readily to the occupying Nazis. But here in 1918 was the story of an extraordinary cipher, and an almost miraculous effort to defeat it.)

ADFGVX had its own significance too as a cipher gambit, for when war came again, and the Germans had moved on to the mechanised Enigma system, there was still a need for cunning hand-cipher systems that could be used in the field, especially by special agents trying to keep themselves hidden. This meant that the principles of ADFGVX and the older system called Playfair were brought into play. Sometimes the way to stay ahead of the enemy was to employ methods that were not used on an industrial scale but had more of a flavour of homespun ingenuity.

33. THE EROTIC BETRAYALS

It was an exotic stage name that gave her a form of immortality: 'Mata Hari' still conjures all sorts of images of sensuous, sultry espionage. Her life became characterised by seduction and whispered words, ingenious wiles and bewitching performances. Mata Hari became a byword for the most ruthless form of spying, in which sex was deployed as a weapon. The truth – as is always the case when the showbusiness mask is dropped – was very different, and in its own way rather tragic. Yet the woman who was Mata Hari did indeed play a role in the spying game of the First World War, and part of that game was the use of cunning encryptions. Sadly, it was the codes that betrayed her in the end.

Her real name was Margarete Gertrud Zelle. Born in the Netherlands in the late nineteenth century, she had fallen in love aged seventeen with an abusive drunkard many years her senior. Rudolph 'John' MacLeod was a Dutch East Indies officer, and with him she forged a life in Java, Indonesia. For her it was a form of hell. She had two children, one of whom died as a baby; then, in 1905, even though it was his own serially adulterous behaviour that was intolerable, MacLeod decided to leave her and also take custody of their daughter. After all this trauma (she eventually got her daughter back), Margarete returned to Holland. But something psychologically had snapped and she left the girl with family as she vanished elsewhere. The next time Margarete Zelle reappeared, she was in Paris, and her name was Mata Hari.

The name was from Java, and it meant 'Eye of Dawn'. In licentious Paris, Mata Hari's stage act was one of erotic dancing, either in skimpy and orientalist costumes or simply naked. But she was not performing in dives. This was an act aimed at the rather more select tiers of society, with the Javanese angle throwing in a measure of apparently respectable anthropology. In other words, the men were not merely ogling, they were also taking cultural notes. Her notoriety opened some very exclusive social doors. When war came to Europe in 1914, Mata Hari's sexual charisma also brought her to the attention of the French secret service, and the German secret service too. Her lover, Captain Vadim Maslov, a Russian serving with the French air force, had been shot down and badly wounded. Zelle, as a Dutch citizen, was free to move between borders and so she asked the French authorities if she might visit his hospital at the front. The authorities permitted this, but only on condition that she agreed to carry out espionage work for France.

Her cover, paradoxically, was her international fame. It was the erotic dancing act that allowed her to continue crossing borders even in a time of war. Who could suspect that such a performer would be concealing secrets? She was hiding in plain sight. And so Mata Hari's spying career began. One plot was for her to try to seduce Wilhelm, the son and heir of Germany's kaiser. But in order to arrange such a fanciful scenario, she had to travel to Madrid to first entrap a German military attaché who could then procure an introduction. But such was her meeting with Major Arnold Kalle that instead she was inveigled into becoming a double agent, trafficking French secrets to Germany as well as German secrets to France. There was no ideology or just cause, it was an auction of financial inducements. On top of this, she must have understood that she was already lost in a terrible maze and perhaps the only way out of it was money. Her life became increasingly messy and into that life came the additional complication of the coded messages.

Part of her espionage career now involved transmitting encoded messages by radio, and these were always going to be intercepted. In a curious twist, a cross-Channel voyage brought her to Falmouth, Cornwall, where she was arrested by the British, apparently mistaken for another female espionage suspect. She was taken to London and interrogated and she told her captors that she was working for the French Secret Service. They were allies, so she was released and allowed to stay at the plush Savoy Hotel. But the engines of her doom were already turning elsewhere. She was betrayed by Major Kalle, who sent enthusiastic messages to his superiors about Agent H-21. He sent these messages in a code that the French interceptors had already learned how to unravel. And it was very plain – for he had done little to disguise the fact – that 'Agent H-21', who had been passing on secrets about figures in French high society, was none other than Mata Hari.

The suspicious French fed Mata Hari information about some Belgian agents, one of whom they also suspected was a double agent working for the Germans. Shortly afterwards, that double agent was executed by the Germans. Was this proof that Mata Hari had betrayed the man to her spymasters? In addition to this, the messages that she was cabling back to Germany were in a cipher that the French had already broken into. Back in Berlin, Mata Hari's spymasters considered her usefulness at an end. She was paid off by means of a cheque, but when she returned to France, the French secret service swooped in and arrested her. This seemed to spark a dreadful hysteria in the authorities. Mata Hari stood accused of betraying the lives of at least 50,000 soldiers in the field, her coded messages blamed directly for battle deaths.

The charge lacked weight but there was a riptide of passion and anger about the carnage that the war against Germany had become, and Mata Hari was the lightning rod for that rage. The fact that her spying had barely conveyed anything but gossip and obvious facts was ignored. She was put on trial in the summer of 1917 and fought hard to be heard, insisting that she was a dancer, not a master cryptologist or a secret agent. She begged her former lover Captain Vadim Maslov to speak up in her defence, but the wounded airman declined. This refusal broke her heart and also probably signalled to her that salvation was impossible.

Painted in the courtroom as an agent who had made a career out of manipulating men for her own dark ends – thus initiating the archetype that still plays out in dramas today – Mata Hari could find no form of words to defend herself adequately. She was found guilty and met her death before a firing squad – it was said that she refused a blindfold so that she could gaze unflinchingly upon her executioners. In a curious way, she was a sacrificial victim. Her life was taken as some form of recompense for the wider bloodshed of the war. The irony was that ciphers, which normally obfuscate and obscure truth and identity, left her standing as plainly as though she was under a dancer's spotlight.

1

MESSAGE FROM THE TRENCHES

Officers in northern France who could speak some French were clearly very valuable in communicating with their French allies. An important number made up of six digits is concealed in the message below. Reading the sentences out loud might help you to crack this code.

HOSTILITIES MUST CEASE.

PLANS FOR AN ARMISTICE ARE ALL SET.

THERE HAS BEEN AN ACCIDENT INVOLVING A LORRY LOAD OF WHEAT.

IN THE MUD, THE VEHICLE SANK.

WE HAVE HAD TO ABANDON THE VAN.

2

BUILDING BRIDGES

Bridges were key strategic points on a Great War map. They were the means by which countries were crossed and regions were linked. Here you need to discover the coded message by uncovering the link word in each instance; one word will complete the first word or phrase and begin the next. When you have completed all ten links, you will have the important message, which can be transferred out in to the field.

1 OUT (_ _ _ _) BACK

2 LEAN (_ _) WARDS

3 HOME (_ _ _ _) SHIP

4 GUILD (_ _ _ _) MARK

5 DOUBLE (_ _ _ _ _) WIND

6 PASS (_ _ _ _) ALL

7 DO (_ _ _ _) STREAM

8 SET (_ _ _ _ _ _) ROOT

9 UP (_ _ _ _) HEART

10 HEALTH (_ _ _ _) FULLY

3

CODENAME

Tom, Dick and Harry are code operators in the trenches. Their code names are Pip, Squeak and Wilfred, which became the nicknames of three World War I campaign medals, the 1914–15 Star, the British War Medal and the Victory Medal. There was also a post-First World War cartoon strip, first seen in the *Sunday Pictorial*, which told of the adventures of Pip, a dog, Squeak, a penguin and Wilfred, a rabbit.

No code operator has the same number of letters in his real name as his code-name.

Two combinations of real and code-name contain the same number of letters.

Which code operator has which codename?

4

MATA HARI

A word square reads the same whether going across or down. Each word doubles up and can be seen twice, as befits someone believed to have been a double agent. Use the listed words to complete two word squares, one including the word MATA, the other including the word HARI.

There is one word that is left over when the grids are completed. Which word is it?

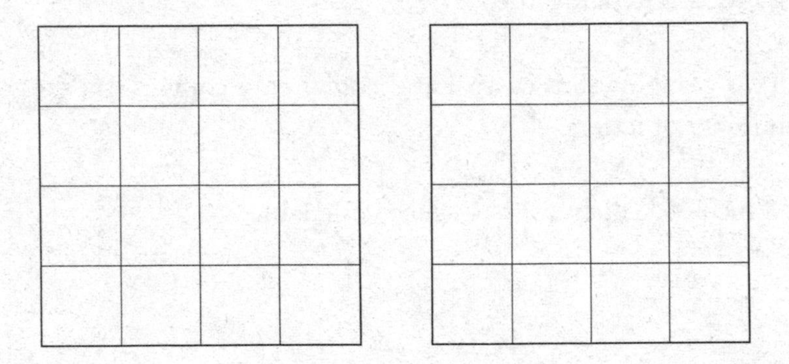

ARID CHAT FILM HARI
IDEA LENT MATA RISK TIDE

5

A CENSUS PUZZLE

Here, below and overleaf are four vintage Edwardian puzzles, devised by Henry Dudeney, and of the type that were relished by Bletchley genius Alfred Dillwyn 'Dilly' Knox.

Mr and Mrs Jorkins have fifteen children, all born at intervals of one year and a half. Miss Ada Jorkins, the eldest, had an objection to state her age to the census man, but she admitted that she was just seven times older than little Johnnie, the youngest of all. What was Ada's age? Do not too hastily assume that you have solved this little poser. You may find that you have made a bad blunder!

6

MOTHER AND DAUGHTER

'Mother, I wish you would give me a bicycle,' said a girl of twelve the other day.

'I do not think that you are old enough yet, my dear,' was the reply. 'When I am only three times as old as you are you shall have one.'

Now, the mother's age is forty-five. When may the young lady expect to receive her present?

7
CHANGING PLACES

The clock face below indicates a little before 42 minutes past 4. The hands will again point at exactly the same spots a little after 23 minutes past 8. In fact, the hands will have changed places. How many times do the hands of a clock change places between three o'clock p.m. and midnight? And out of all the pairs of times indicated by these changes, what is the exact time when the minute hand will be nearest to the point IX?

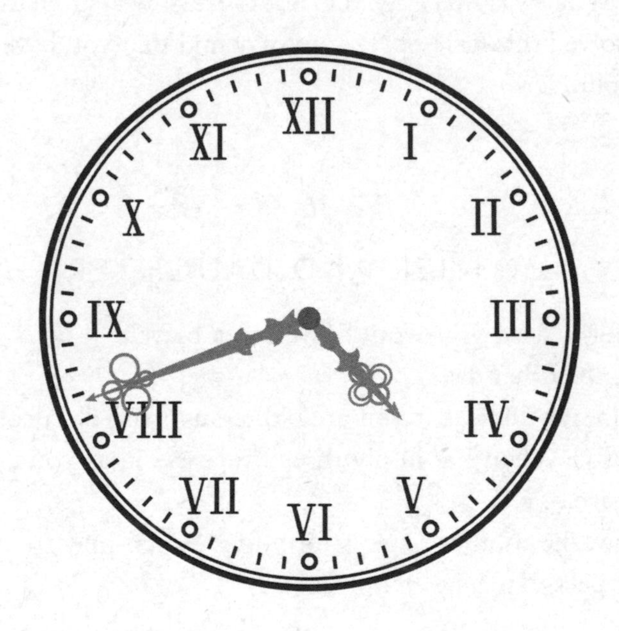

8

CURIOUS NUMBERS

The number 48 has the peculiarity that if you add 1 to it the result is a square number (49, the square of 7), and if you add 1 to its half, you also get a square number (25, the squre of 5). Now, there is no limit to the numbers that have this peculiarity, and it is an interesting puzzle to find three more of them – the smallest possible numbers. What are they?

9

BEESWAX

Here, and overleaf, are three vintage cryptological puzzles devised over 100 years ago by Henry Dudeney, at around the time of the Room 40 codebreakers...

The word BEESWAX represents a number in a criminal's secret code, but the police had no clue until they discovered among his papers the following sum:

E A S E B S B S X

B P W W K S E T Q

K P E P W E K K Q

The detectives assumed that it was an addition sum and utterly failed to solve it. Then one man hit on the brilliant idea that perhaps it was a case of subtraction. This proved to be correct, and by substituting a different figure for each letter, so that it worked out correctly, they obtained the secret code.

What number does BEESWAX represent?

10

WRONG TO RIGHT

'Two wrongs don't make a right,' said somebody at the breakfast table.

'I am not so sure about that,' Colonel Crackham remarked. 'Take this as an example. Each letter represents a different digit, and no 0 is allowed.'

```
W R O N G +
W R O N G
---------
R I G H T
```

'If you substitute correct figures the little addition sum will work correctly. There are several ways of doing it.'

11

THE CONSPIRATORS' CODE

Two conspirators had a secret code. Their letters sometimes contained little arithmetical sums relating to some quite plausible discussion, and having an entirely innocent appearance. But in their code each of the ten digits represented a different letter of the alphabet. Thus, on one occasion, there was a little sum in simple addition which, when the letters were substituted for the figures, read as follows:

```
    F L Y
    F O R
  Y O U R
  -------
  L I F E
```

It will be found an interesting puzzle to reconstruct the addition sum with the help of the clue that I and O stand for the figures 1 and 0 respectively.

12

ANOTHER POLYBIUS SQUARE!

Here, once more, is an encryption system of some elegance and a little subtlety. Pioneered by the ancient Greeks (and popularised by the scholar Polybius), this enabled coded messages to be broken down into small co-ordinates, which were easier to convey. More than this: the 25 Greek letters were rendered into numbers. In today's version, 26 letters have to be crammed into the 5x5 frame – so 'I' and 'J' share a square.

How to unravel the codes: it is essentially a question of reading off co-ordinates. Each two-digit number represents one letter. The first of the two numbers indicates the row (vertical); the second number indicates the column (horizontal). For the number 34, for instance, track down to row three, then move along horizontally to number 4; there your letter will be.

And the same system applies when you wish to devise codes of your own.

	1	2	3	4	5
1	A	B	C	D	E
2	F	G	H	I/J	K
3	L	M	N	O	P
4	Q	R	S	T	U
5	V	W	X	Y	Z

1 45 33 14 15 42. 43 24 15 22 15.
43 15 33 14 21 42 15 43 23
13 34 23 34 42 44 43

2 52 15 32 15 15 44 21 34 42
12 11 44 44 31 15 34 33 44 23 15
35 31 11 24 33

3 44 23 15 11 42 42 34 52 43 11 42 15
44 24 35 35 15 14 52 24 44 23
35 34 24 43 34 33

4 12 15 52 11 42 15 44 23 15 32 11 55 15
34 21 25 33 34 43 43 34 43

5 44 23 15 43 35 11 42 44 11 33 43 11 42 15
32 11 43 43 24 33 22 24 33 44 23 15
33 34 42 44 23

CHAPTER TEN

A CONTINENT OF CODES

In which we trace how Europe's long, bloody history of sieges and atrocities led to the rise of dedicated secret codebreaking departments.

34. THE INVISIBLE CHAMBERS

A continent roaring with fire and war and blood, savage battles and merciless persecutions, where powerful monarchs, emperors and dukes engaged in deathly struggles over religion and statehood. Europe of the 1600s was in some eyes a vision of the apocalypse. And in that century it was the French under Louis XIII who led the way in deep-state cryptography with the Cabinet Noir, or Black Chamber, where codebreaking was turned into a secret underground industry. This was an age when messages and their interpretations could decide the fates of individuals or cities.

The Black Chamber was an early form of intercept station. Rather than radio signals, the operatives were working on the written communications that were passing between embassies and more exalted private citizens, between government officials and the great religious houses. The business of breaking seals on letters, undetected, was itself a matter of some skill, before the business of codebreaking the letter could begin. Those who worked in the Cabinet Noir were experts at making it look as though correspondence had been untouched. In addition to this, they were accustomed to working quickly by either breaking the cipher or transcribing the coded letters before sending them on. All delays in delivery had to be kept to the bare minimum so that none might suspect that their private business was being monitored. (And indeed, the Cabinet Noir was still going strong in Napoleon Bonaparte's day, when each morning he would be presented with a leather file of decrypted letters from the secret department.)

The extraordinary storm of the Thirty Years War throughout the Holy Roman Empire, and sucking in France and Spain among others in the middle years of the 1600s, had summoned forth its own codes. Only recently, a cache of enciphered seventeenth-century correspondence was unearthed in Vienna and modern-day codebreaking sleuths have examined copies and have been eagerly suggesting solutions some four hundred years on. Vienna itself hosted what was to become renowned as one of the greatest Black Chambers, deep wthin the Hofburg Imperial Palace. It was called the Geheime Kabinettskanzlei (or Secret Cabinet Office). And it was a chillingly efficient operation that was already at work before the first rays of sunlight met the spires of Vienna each and every morning. Its chief business was diplomatic traffic (just as 300 years later there were special codebreaking departments within the Bletchley operation focusing upon diplomatic cables). Among these desks and candles and lamps were clerks who would carefully warm and melt the wax seals on letters without causing any blemish or damage to the paper; then there were those who assiduously copied out the contents of each letter, followed by the experts who could bind up the letters once more by brilliantly recreating the elaborately stamped wax seals (all with their own unique motifs). The result was that the proper recipients had no idea that others had been delving into their private conversations.

This was not only a successful espionage operation in its own right – though it is difficult to imagine how ambassadors and diplomats could have been so naive as to not think that something of the sort was happening – but the Vienna Black Chamber also became an exporter of the gravest secrets for money. In essence, this was the first instance of government codebreaking being privatised to turn a profit. The juiciest scandals and secrets were sold on to agents of the French throne (in the pre-revolutionary days). There were some wise aristocrats and nobles who were concerned that Black Chambers anywhere might be spying upon them, and there were examples of amazingly elaborate origami-style ruses involving unique and complex ways of folding paper that could not be easily replicated. Thus, the slightest crease out of place would alert any recipient of such to the danger of intelligence having been stolen. There were feints too, like the false envelope. A vitally important letter would be addressed to someone of the least significance, in the most ordinary street. Such addressees would never have their mail violated because they were of too little importance. But these addressees were agents in their own right, paid middlemen. They would receive the unopened letters which they would then deliver to the true recipient by more direct and safer means.

There were other ruses too – for example, bypassing the straightforward mail system altogether by concealing letters within freshly baked pies or loaves, which were delivered to the officials or nobles concerned. Secret though the Black

Chamber of Vienna was, everyone had too sharp a sense of what might be possible if the state decided that a noble person was worth monitoring. Some found out in the most bitter way that their secrets were undone, and the monarch Louis XV of France was one. In 1774, a bumper package was delivered to him which contained, rather to his astonishment, the most confidential orders given to spies by the King of Prussia, and also, to his dismay, copies of enciphered – and subsequently decoded – letters that he himself had sent. The source of this haul was a secretary to the French ambassador in Vienna. This secretary, apprised of the invisible trade of the Black Chamber had set out to ascertain the reach of its intelligence-gathering. One night, as the bells of midnight sounded across Vienna and spies watched glittering-eyed from impenetrable shadows, the ambassador's secretary made contact with one of the Black Chamber codebreakers and handed over the handsome sum of 1,000 ducats. The money yielded up the intelligence that the horrified king was now surveying.

And the lasting significance of this and other Black Chambers in other European cities was that codebreaking had not only been enveloped into the bosom of the state but it had also evolved its own systems and methods of keeping out of the light as ciphers were broken on a brilliantly efficient production line.

35. THE PAPAL SECRET SERVICE

The villages and towns of medieval Europe were frequently visited with catastrophe. Harvests would fail and hunger would stalk the land, plague would take a hold and extinguish countless lives. Faith might not have provided physical comfort in this world of pain, but it did offer the certainty of the life beyond.

Yet religious conflict too was one of the great scourges that periodically caused blood to flow in torrents. The late 1300s saw the Great Schism of the Roman Catholic Church, and it was in this period that succeeding popes began to use cunning cryptography. Their everyday communications – even on matters of delicate biblical dispute – could be dangerous. So they had to be obscured. Then, in later centuries, as the Vatican grew and papal influence extended to countries around the world, the need for secrecy increased. So much so that Vatican encryption became a minor industry in its own right.

In 1379, Clement VII inaugurated the new age of ciphers, via his secretary Gabrieli di Lavinde, who devised a system which used both substituted letters and code lists in which whole words and names were substituted. Clement VII was a pope who had fled Rome to Avignon, thereby becoming an 'antipope'. The encrypted system enabled Clement to correspond with clerical allies without being detected.

By the 1500s, and the storm of the Reformation, cryptography was embedded deep in the Vatican. Not least because it was understood that diplomatic letters sent to countries like England were being diverted, opened and read. It was in 1555 that the Church acquired its own full-time official cipher secretary. But as popes came and went, this job itself became the focus of furious conflict and internal politics. A young man called Matteo Argenti, the nephew of a previous office holder, Giovanni Battista, came to the fore in 1591, under Pope Gregory. He had an eye for devising distinguished and baffling ciphers. Just a few years later, this crucial figure was ejected in another of the endless factional battles surrounding the papacy, but his legacy was rather more permanent. It was an elegant cipher system involving letters, numbers and tricks such as merging letters that always followed one another ('q' and 'u' for example) into one encoded letter. There was also the trick of bunching all words in the message together so that no one could tell how long any one individual word was supposed to be.

Thanks to these and other efforts, the cryptology department of the Vatican became world-leading for a time. Eventually, different nations would start their own dedicated departments, or 'Black Chambers' (as seen on page 221). By the nineteenth century, that sense of innovation had long disappeared. But the pope's codes were to have integral importance once again in the twentieth century, so much so that they were attacked from all sides. In the First World War, this was because the Catholic Church held so much sway across such vast swathes of continents, thus it was always necessary to know whenever the pope might be making gestures of peace with any of the combatants.

Come the Second World War and the struggle against the Nazis, and this interest intensified violently. Germany had a residually strong Catholic tradition, which Hitler was anxious to suppress at every opportunity. He was the Führer and he could not allow his authority to be shared with that of the pope in Rome. At the same time, it was not politic to completely ban or suppress the Church, as faith ran too deep and too strong. But the Nazi distrust of the Vatican was pulsating and intense, and the mastery of the papal ciphers was one of the priorities of Hitler's codebreakers. They had head starts, having secretly acquired various Vatican codebooks, giving them keys. When the blaze of war began, there was an operational urgency about the Nazi espionage. As well as providing succour for people in Nazi-invaded territories such as Poland, the Church was also thought to have strong links with local intelligence services. The activities of resistance fighters, and their aims, could have been relayed to the Church through these cipher systems.

And the Vatican had been keeping a close eye on Hitler since his days as a beer-hall revolutionary in the 1920s. Even from the start, it detected that he and his followers could pose the most terrible threat to Catholics everywhere. He was the subject of encrypted communications years before he ascended to power. The Nazis were not alone in seeking to penetrate these codes. The British were at it as well (on the grounds that all intelligence had its uses), as indeed were the Italians (the Vatican quite naturally kept its cryptology efforts completely separate from the fascist regime of Mussolini). In addition to this, Stalin's Russia was implacably hostile towards the Vatican, even though it made diplomatic noises to the contrary. The reasoning was similar to that of Hitler. The pope stood outside of the laws of the Communist regime and he commanded a loyalty among worshippers that should have belonged exclusively to Stalin. The Russian Orthodox Church was tolerated in this time of totalitarianism because it kept the prospect of heaven within one state. But to have masses of Catholic worshippers in already recalcitrant states such as Poland carried the threat of serious political opposition.

For these reasons, Vatican ciphers were as avidly attacked as those from any national embassy. The pope transcended borders and the totalitarians had to know who the Vatican was in contact with. Happily, a great many of its encryptions refused to yield to unravelling and in war (and the Cold War), with limited time and resources, codebreakers often had to turn to more urgent messages. In many ways, the Roman Catholic Church was the European seat of modern cryptography, its intricacies having been mapped out in exquisite Renaissance gardens. But the vital importance of those codes stretched deep into the darkness of the twentieth century.

36. THE REAL DA VINCI CODES

A hill-top chapel overlooking a rich valley and distant mountains, deep in France. An excavation and two mysterious encoded parchments. Legends that date back over two thousand years and a secret that could rewrite the whole of western history. Of all the ciphers that have ever existed, few have had the power to feature both in fiction and reality. Next to none have exerted a grip on the global imagination quite as much as the curious writings found within the church of a tiny village near Toulouse called Rennes-le-Château.

The codes themselves seem real enough, but the industry that they have inspired across the last few decades has at times appeared almost surreal. These ciphers were apparently discovered by a turn-of-the-century priest called Bérenger Saunière, who subsequently and puzzlingly found enormous sums of money with which to renovate the Rennes-le-Château estate. These ciphers were then written about in the 1970s and 1980s by a screenwriter (with *Doctor Who* among his credits) called Henry Lincoln. His non-fiction work gave a thrilling cohesion to legends of treasure and a long-buried secret that could never be whispered of (spoiler alert: the secret will be revealed within the next few paragraphs!). This in turn inspired fiction authors as diverse as Umberto Eco (with his mighty philosophical novel *Foucault's Pendulum*) and the rather less philosophical Dan Brown in *The Da Vinci Code*. What is this extraordinary cipher that inspired and continues to inspire so many?

In fact, there are two parchments, one larger than the other. They are written in Latin, though there are curious anomalies and symbols dotted around both documents that for some time resisted the efforts of various codebreakers. They are suffused with biblical imagery. And it was Henry Lincoln, with several colleagues, who began to piece the ciphers together and from them extrapolate a web of conspiracies that stretched back to Jerusalem 2,000 years ago, linking in the French Merovingian dynasty, the Cathars and various secret hermetic religious groups.

This most beguiling of code mysteries really started to take off in the 1960s, when the new owner of Rennes-le-Château began looking into the curious symbolism of the renovated chapel, and into the life of its distinctly unconventional priest, Bérenger Saunière. The legend was that during excavation of the chapel, Saunière had discovered that one of the pillars was hollow and that it contained a vast amount of gold pieces buried by long-dead nobles. Part of that haul was used to turn Rennes-le-Château into a dazzling destination. The chapel itself was adorned with curious and fearsome symbols, including a demonic figure near the door with the legend 'This is a terrible place' intended as a form of cryptic warning, and stylised reliefs of scenes from the Gospels.

Also in that hollow pillar were the two parchments. Some of the more inexplicable writing on them could apparently be put through a Vigenère Square (see page 45) with the use of a phrase key derived from one of the tombstones in the cemetery. Saunière himself later came to find himself in disgrace, not for any of these stories but for the religious offence of selling masses. He was evicted from the parish by the local bishop and went on to lead a life of more conventional priestly poverty. But this was material for later conspiracies. Did the coded parchments he had stumbled upon contain too great and too destabilising a secret for the Church to absorb? Did these ciphers contain a revelation that in fact could overthrow the entire basis of Christianity? By the time Henry Lincoln was on the case in the 1970s, making documentaries for the BBC and devoting himself to esoteric research, the contours of the secret seemed to be coming into focus. References in the parchments to the Holy Grail were especially gripping. There had long been a legend throughout Europe that when Christ was on the cross, the cup from which he had drunk at the Last Supper, and into which some of his blood now ran was taken by Joseph of Arimathea (and, according to some legends, ended up in Glastonbury in Somerset).

Lincoln and his colleagues were alive to the power of metaphor within these ciphers too, and so it was not just the letters that were ravelled but also their literal meaning. The site of Rennes-le-Château had been associated with ancient Merovingian royalty, dating back 1,500 years. Was it possible that the Holy Grail itself was a metaphor? Other clues within the parchment pointed to Mary Magdalene. And, brilliantly, the meaning of it all sharpened further before Lincoln's eyes. The Holy Grail was not a cup, it was the womb of Mary Magdalene. The thesis was that Jesus had married Mary Magdalene and had had children with her. The Holy Grail was his bloodline. After his crucifixion, Mary and family had emigrated to France and there, over successive generations, marriages were made with nobles. Therefore the Merovingian dynasty was directly descended from Christ himself.

Here, the thesis became even more flavourful. It was said that if the Merovingians were ever restored, as various hermetic religious orders (and apparently the Knights Templar) hoped, then the entire Catholic Church would be threatened. For what standing would the pope have in the face of the direct descendants of Jesus? Thus it was that various sinister powers did their best to keep the secret suppressed.

The book that Henry Lincoln wrote about all this, with Michael Baigent and Richard Leigh, was, for a time, to be found in millions of sitting rooms across the land (my parents had a copy). It went on to inspire not only Dan Brown, and his bestselling fiction but also, rather more intellectually, the philosopher Umberto Eco, who took the story and treated it as a fragment of an even *larger* global conspiracy, involving yet more ciphers and codes. Although it is not the simplest of reads, his 1989 entertainment *Foucault's Pendulum* is in many ways a soaring love letter to secret encryptions. It is a thriller in which three Italian publishers are approached by a mysterious colonel who claims to have uncovered a plot by the Knights Templar to take over the world. The publishers are long accustomed to meeting conspiracy theorists – such people always want to make it into print. But then the colonel swiftly thereafter disappears. The youngest publisher, Causabon, has himself an interest in the Templars, and his colleagues Belbo and Diotavelli have a fondness for conspiracy theories. They set out on what they think is an amusing quest to uncover this secret global plot, tracing the history of all conspiracies from the time of Christ onwards. But as they do so, they find, creepily, that there are real-life figures who are taking a malign interest in their work, and as they work their way closer through the codes and secrets of recent centuries, from Dr Dee to the Rosicrucian sect, it becomes clear that these shadowy others are willing to kill to protect the ultimate secret.

Part of the joy of this densely detailed story is that Eco has patterned it with ciphers and encryptions. The publishers use a computer they call Abulafia and feed it conspiracy theory phrases that the machine then jumbles and randomises, creating brand-new conspiracy theories. But this in turn is actually tangentially related to an Ancient Hebrew form of encryption, called the Notarikon.

This was a means by which phrases and sentences from the Talmud would be fragmented by rabbis in such a way that they produced new phrases, so revealing a hidden layer of meaning beneath the words on the page. In *Foucault's Pendulum*, it is the older publisher Diotavelli who feels the first pangs of unease about playing with language so lightly. He believes words can have a direct physical effect upon both the body and the soul, and that by turning them inside out to discover hidden secrets the body and the mind could end up being distorted too.

And the global plot itself is a parable about the dangers of interpretation. It dizzyingly involves the titular pendulum in a Paris museum, a nineteenth-century means of proving that the earth is rotating; the story leaps from there to energy lines beneath the earth's surface, with a line that points, above the hinge of Foucault's pendulum and beyond into space, to the very centre of the universe. Codebreaking is about detecting patterns, and using those patterns as a principle. Likewise, the world itself might be interpreted with patterns being spotted everywhere. But see too many patterns, or overlay them in the wrong places, and the foundations of sanity can begin to slip.

1

CROSSING PLACES

The towns and cities listed below have had the vowels in their names removed. When you have worked out the coded names, slot them in their correct places reading DOWN in the grid. The new shaded word going across meets at the centre with the shaded word going down.

1 M S S N

2 B D J Z

3 T L L N N

4 T R C H T

5 L R D S

6 G R N D

7 B L G N

1	2	3	4	5	6	7

2

EURO TOUR

Take the European Grand Tour and find the countries and principalities in the codes below. There is no starter to guide you on your journey. Singlehandedly, you must find your way to the EIGHT different destinations.

A 1 2 3 4 5 2 6

B 5 2 6 15 8 8 5

C 6 1 2 7 5 8 9

D 10 11 1 6 1 2

E 10 12 5 13 2

F 1 10 14 15 2 13 5

G 7 15 2 5 16 15

H 12 15 4 5 2 6

3

MYSTERY IDENTITY

Look at the sentences below. Hidden in them are words and names which when joined together will spell out a message, one which can be interpreted the world over. Can you crack this code? The speaker recognises who he or she is talking to but has other problems with recollection.

1. My sister is unwell, I'm afraid.
2. I found singing so much fun I formed my own choir.
3. Here is the book I lost in the library, or so I thought.
4. Evie chose a new dress for this special occasion.
5. He turned round and said, 'Show his keys to me.'
6. The huntsman shot elks in the deep forest.
7. In the drama, the scene of pathos carried so much emotion.
8. I spoke with Ralph after the meeting had ended so abruptly.
9. Visiting the museum, I keep being reminded of days gone by.
10. The secret cell is seen in diagrams on the wall.

4

FLY THE FLAG

Fly the flag by cracking the code. Solve the cryptic clues and work out which word with a nautical flavour belongs to which flag. Only SIX different letters are used.

1. Observes, we hear, on the oceans
2. Seat overturned in this direction
3. Flagpole assembled as one body by the sound of it
4. Rodent returns sailor
5. Schoolteacher in charge of a ship
6. Is this admiral at the back?
7. Guide bullock
8. High regard for one in authority
9. Vessel at sea – not in the kitchen
10. Naval pal with bad table manners?

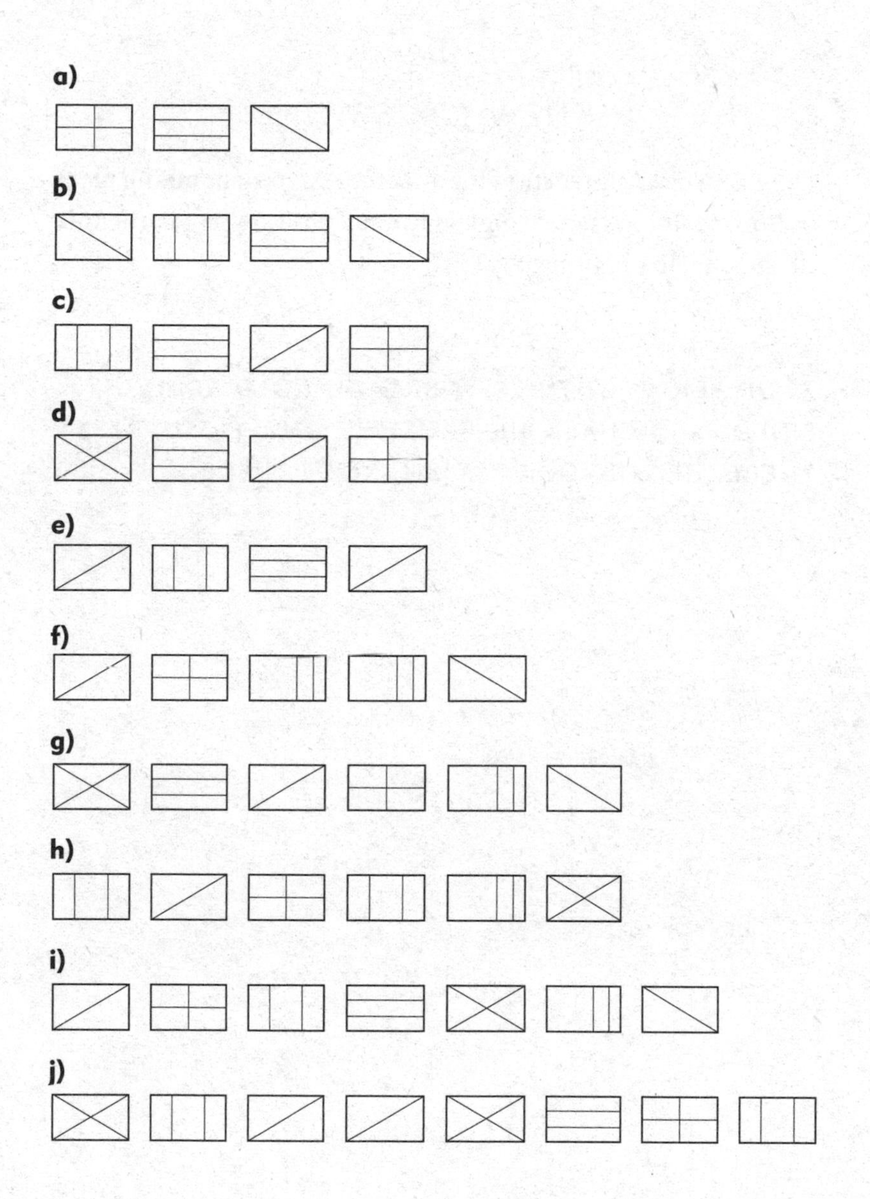
a)
b)
c)
d)
e)
f)
g)
h)
i)
j)

5

KNEW OMISSION

Here is a coded instruction for a secret assignment taking place in Europe. It looks meaningless but it isn't! Looking at the title of the puzzle might help.

TOUR PLAIN. GOT FEAST. BREACH COLD BRANCH. FINED FRED CHASE BIN BRUSHES. PAMPERS CARE THERE. HEARD FOUR PARISH, KNOT NIECE.

CHAPTER ELEVEN

ENIGMA AND ITS VARIATIONS

In which we examine the astonishing ways in which the women and men of Bletchley Park left such a profound mark upon the history of the Second World War.

37. THE WOMAN WHO SANK THE NAVY

The night skies above Britain in the autumn of 1940 were heavy with German bombers and those same skies glowed with an uncanny ferocity as the flames of the Blitz consumed dockland warehouses and residential streets. During this time there was a young woman who was a gifted linguist as well as a literary scholar studying at the capital's university. She and her fellow students were evacuated out to a rather more remote academic institution, the University of Aberystwyth on the west coast of Wales. There, the young woman tried to apply herself to the study of eighteenth-century German Romantic poetry. But, she thought to herself, how could she focus on this while in the present day young German men were dropping bombs on British cities? It was not long before she was recruited to one of the war's most intensely secret establishments. And it was there that nineteen-year-old Mavis Lever was to make an extraordinary contribution to the course of that war.

The establishment was Bletchley Park. Her initiation had been to make sense of a translated Morse code message that no one else could fathom, consisting of the letters S T G O C H. Was it an acronym? Was it a non-existent 'St Goch'? But the pure light of inspiration hit Mavis quickly. This was a message about South America. The letters in fact were an abbreviation for Santiago, Chile. Her speed and aptitude showed the authorities that they had a perfect, and brilliantly talented, new recruit.

Like Mavis, the codebreakers of the Government Code and Cipher School had been evacuated from London, anticipating the bombing campaign. The Park comprised a (slightly architecturally eccentric) late-Victorian mansion in the north of Buckinghamshire. By the house was a lake, upon which lived some very short-tempered geese. And in front of the house was a gentle lawn, upon which, on some summer's evenings, young codebreakers could be seen performing Highland dances, with bagpipes wailing out of a gramophone. It was into this bewildering world that the sharp Mavis Lever was thrust.

She was sent to work in a department known as The Cottage. It was an outbuilding next to the house by the old stables. In charge here was Alfred Dillwyn 'Dilly' Knox, a wayward classicist who had contributed to the Zimmermann Telegram decryption triumph back in 1917 (see page 197). In the intervening years, he had taken the most intense interest in the new electric enciphering machine that had been brought into use by the German and Italian military. A compact wonder called Enigma, this machine looked a little like a portable typewriter, and it could be used in any theatre of war, generating some 178 million potential combinations of new codes. The Germans and their associates assumed, not unreasonably, that the machine's encryptions were completely unbreakable. What human brain could begin to process that terrifying number of possibilities?

Mavis Lever had moved from one kind of university to another, and her tutor in the esoteric arts of codebreaking was one of the few men anywhere who had begun to see how the Enigma might be outwitted. Dilly Knox had been given a formidable push by three Polish mathematicians (see page 253) who had devised ways to crowbar into it in the early 1930s. The methods were all extraordinarily arcane. On Mavis Lever's first day in The Cottage, Knox placed before her some Enigma-encrypted text, then said, 'Hullo, we're breaking machines here – have you got a pencil? – here, have a go!' and she exclaimed that it 'was all Greek' to her. Knox delightedly replied, 'Oh, if only it was!'

In terms of teaching, Knox also talked to her about Lewis Carroll and the author's lateral-thinking linguistic and mathematical riddles. He explained that it was about seeing the world from new angles and dimensions. Knox had also worked upon his own extraordinarily abstruse means of bringing forth sense from Enigma. He termed it 'rodding' and it was a linguistic system that imagined all the code wheels of the Enigma, with all the letters on them turning, and expressed that in vast grids. The long parallel lines of letters that might correspond. Mavis Lever had the intellectual spirit to make the leap into this realm, and to gradually make it a form of second nature.

She was joined in this by another young woman, Margaret Rock. Internal Bletchley Park gossip suggested that Knox employed beautiful young women for lascivious reasons, and they were even termed 'Dilly's Fillies'. But this was intensely insulting. Lever and Rock were there because Knox believed that women were better codebreakers. Their lightning intellects had a brilliant flexibility. Knox had used his rodding system to make a little headway into the Italian Enigma system. Mavis Lever was to run with the system and in doing so produce the means by which a great and famous victory in the real world could be won. In 1941, an enciphered message from the Italian Navy landed on her desk, only the tiniest fragments of which could be fathomed. She set to work on it over the space of three days and nights, scarcely stopping for tea or meals or sleep. Eventually its true meaning yielded itself to her: 'Today's the day minus three.' This was about a projected Italian naval attack in the Mediterranean. Intelligence was swiftly passed to the Admiralty.

What then followed was a terrific naval skirmish called the Battle of Cape Matapan, where the Italian vessels were comprehensively beaten. It was a moment of elation for Admiral Cunningham and the British Navy, but it was also a cause for euphoria at Bletchley Park. A nineteen-year-old had influenced the course of a battle fought over a thousand miles away. And Mavis Lever was to do very much more. She set to work on penetrating the Enigma codes used by the Abwehr, the German secret service. Again, after a period of great intensity, she managed to prise her way into the system. This was to prove beyond value, because when British intelligence efforts to misdirect the Nazis with false information were underway, it was possible to see, through the secret service messages, how far these schemes were working. And arguably the greatest of these successes was the hinge of victory in the entire war.

In order to disguise the forthcoming invasion of the French beaches, known as D-Day, British intelligence had been seeding fake intelligence throughout the Continent about where the Allied forces were planning to make their landings. Thanks to the unravelling of the Abwehr secret service codes, the Allies could see that the Nazi regime had fallen for their bluff about the incursion coming further north in France. D-Day had the crucial element of surprise, without which, as another Bletchley whizz Harry Hinsley pointed out, it might have ended in disaster with the Allies being thrown back into the sea. And that, in turn, might have prolonged the war by another two or three years, costing uncountable lives. And at the eye of these epoch-forming events, Mavis Lever kept her cool so brilliantly that some American codebreakers, who joined up with Bletchley Park, could only gape at her with wonder. Who was this prodigy who was taking the lead in among departments of middle-aged men? The US cryptologists had never seen anything quite so modern.

It was also at Bletchley where, in 1942, Mavis Lever met fellow codebreaker Keith Batey, who was also an undergraduate student. They fell in love (Dilly Knox cannily observing their courtship) and married. After the war, it was generally expected that women would return to the home and be housewives, an intolerable fate for one who had been at the centre of the world's shifting tides of war. But new challenges awaited Mavis Batey when a new branch of intellectual enquiry led her into landscape history. She was a pioneer in decoding the lost secrets of Britain's moors, commons, fields and gardens and wrote many books. And all the while she was required by the Official Secrets Act to never breathe a word about her former life.

Happily, by the mid 1990s, the secrecy lifted and she and her husband were at last free to communicate, with sly good humour, the impossible rigour of the codebreaking work they had done. What also came to light were some scraps of poetry from her former boss, Dilly Knox, who on one occasion wrote of the Nazis having been thwarted in part by 'English lassies rustling papers through the sodden Bletchley day'.

38. THE MOST IMPORTANT HUT IN BRITAIN

A large hut, painted pale green so that it might to an extent blend in with the grass on which it stood; and contained within it, a narrow central corridor off which ran a series of individual rooms. These rooms were sparsely furnished with desks, chairs, occasionally some radio equipment, blackboards and walls with pinned maps. Because of the threat of air raids, the windows were covered with black-out tape, which gave the hut a permanent twilight feel. In the summer, it was stifling, and in the winter, heat was supposed to come from a stove towards the back of the hut, but the fumes were more noticeable than the warmth. In this supremely utilitarian setting, one of the abiding miracles of the Second World War was performed. It was so secret that not even the personnel in similar huts nearby knew the details.

These were the structures that were dotted around the eccentric country house of Bletchley Park. Inside Hut 8 was a team dedicated to breaking into the secret ciphers of the German Navy. Of all the Enigma-machine propositions, the naval codes and their operation were particularly tight and disciplined. There were those at Bletchley Park at the start of the war who doubted whether these particular codes could ever be broken. But failure to do so could tip the balance of the conflict and give the Nazis a stranglehold on Britain that it would not be able to escape.

Yet the team that was installed in Hut 8 was in so many ways exceptional. Headed up by the brilliant mathematician and philosopher Alan Turing, who was still only in his late twenties, there was also the young and talented Joan Clarke (another mathematician) and Conel Hugh O'Donel Alexander (a phenomenal chess player who, in the years to come, would take on Soviet grandmasters in extraordinary Cold War contests). Also among these fizzing brains were architecture enthusiast Rosalind Hudson and young businessman Rolf Noskwith, originally of German-Jewish heritage, but who passed all clearances when interviewed by the polymath novelist C. P. Snow. They were aware that they were facing an apparently impossible challenge, but it was the youthful Turing who maintained faith that the problem was soluble.

The pressure, however, was intense. As the war ramped up after the summer of 1940, German U-boats began stalking British shipping in the wilds of the Atlantic. Britain's supply lines for grain, cereal, fresh equipment and other necessities were vulnerable and slender. Shipping would travel in convoys, with naval vessels alongside for protection. But there was in truth little that could protect them when the U-boat 'wolfpacks', as they were known, got a scent of their blood. The U-boats could silently and invisibly lurk in the waters beneath and their torpedoes would punch through hulls, causing fires, explosions and entire ships to buckle and fragment before finally sinking into the unfeeling depths. The crews of the convoys, both naval and merchant shipping, lived with a constant, acute sense of mortality and vulnerability out in those icy seas.

If the German U-boat codes could be broken then a huge advantage would be gained. Intelligence concerning submarine positions and routes could be secretly digested and precious convoys could elude them by changing course, saving countless lives. This was the weight that lay upon the shoulders of the team of Hut 8. Turing, together with fellow senior codebreaker Gordon Welchman, had developed the bombe machines that came into service in the summer of 1940. These hulking wardrobe-sized contraptions, filled with rotating drums, could motor their way through code combinations at a speed no human could contemplate. By themselves they could not crack codes; they only worked with ciphers that had been partially prised open or guessed at, and then the machine could chew over the rest.

And when it came to the naval ciphers, there was invaluable head-start help from the Royal Navy. The incredible capture of *U-boat 110* in the spring of 1941, a few hundred miles north-west of Ireland, enabled a boarding party to raid it and among the plunder were codebooks which were spirited back to Bletchley under conditions of utmost secrecy. Code settings were changed every day, but this was an indication of the sorts of systems that were being used. To study them, Turing utilised a fabulously complex technique termed 'Banburismus', involving perforated sheets laid upon a light source. It was a form of dynamic logic diagram involving trial and elimination.

But the German Navy, under Admiral Dönitz, was the only branch of the German military that harboured suspicions that their codes may have been broken. Therefore Dönitz ordered extra security features on Enigma. This meant that instead of having three rotor wheels of letters, turning electrically to produce new code combinations, there were now four rotors. This increased the potential number of combinations by an unfathomable degree. And it had the immediate effect of bringing the work of Hut 8 to a juddering halt. It could not have happened at a worse point. More than ever, Britain needed those supply ships, and more regularly than ever, U-boats in their wolfpacks were bringing death to the convoys and their crews. It was down to the extraordinary courage of three British sailors, out on those hostile waters, that a reversal was made possible.

A German submarine, the *U-559*, had been hit and was on the point of plunging forever beneath the waves. Looking on from HMS *Petard*, twenty-two-year-old Able Seaman Colin Grazier and twenty-nine-year-old First Lieutenant Tony Fasson volunteered to swim to the U-boat to retrieve any material they could before it was lost. They stripped naked and swam across, and followed in a boat, to carry back the material they retrieved, by Able Seaman Tommy Brown. The bravery required was immeasurable. The submarine was about to sink at any moment. Fasson and Grazier descended into the murky darkness of its interior, found a wodge of material, including Enigma codebooks, which they extracted and passed to the boat above before continuing their search. And for this they gave their lives – the submarine sank before they could begin to try to get back to the boat, pulling both Grazier and Fasson into the cold, drowning blackness. For many years, thanks to official secrecy around Enigma, their sacrifice could not even be mentioned. Thankfully, they have now been properly memorialised.

The codebooks they retrieved brought Hut 8 back into action and in doing so changed the course of the Battle of the Atlantic. The wolfpacks that had once stalked the waters now found that they themselves were being stalked by naval destroyers. By this time, Alan Turing had been moved sideways at Bletchley, as it was felt that his key talents were not administrative. But the woman who was briefly his fiancée, Joan Clarke, did stay on. Indeed, her contributions to codebreaking outlasted most of her contemporaries. After the war, when Bletchley Park regenerated into the new codebreaking department of GCHQ, she carried on working there and did not retire until 1977.

But all the men and women in that hut in 1942 had experienced a never-to-be-repeated pressure. By turning the course of the naval war, they laid the foundation for the victory to come. The naval codes and the courage of the sailors who made their breaking possible were the most astounding coup.

39. THE VISION IN THE COALS

One of the most beguiling aspects of the Bletchley codebreaking story is the contrast between mighty military forces operating in theatres of war across the globe and the obscure domesticity of the young people whose ingenuity was helping to shape events. In the spring of 1940, Bletchley personnel, many recruited directly from university, had yet to find their feet. In these early days, they had no miracle machines to aid them; those were still being developed by Alan Turing and Gordon Welchman. Nor did they have very much to go on when it came to the nature of the Enigma machine itself, save what could be gleaned from the several models that the Park had managed to acquire. But from these unpromising prospects came an intellectual feat that combined mathematical brilliance with acute psychology, a triumph both of logic and searing human intuition.

It was the work of a young man called John Herivel. His achievement arguably did much to save Britain in the darkest days of 1940. Yet that achievement had to remain an intense secret for decades after the war, leading to one of the most poignant and sad of the Bletchley testimonies. But now that John Herivel and his work are honoured widely, including on the official website of GCHQ, it is also worth exploring the extraordinary circumstances in which this codebreaker influenced the tides of history. All it took was a lodging house in Bletchley, a small, cosy front parlour, a coal fire and a comfortable chair before it. Here was where Herivel had an insight that was Shakespearean in its depth and simplicity.

At its inception, Bletchley Park was already a new kind of codebreaking establishment. While its senior expertise was still provided by many figures who had been in the First World War cryptography department Room 40, including director Alastair Denniston, Dilly Knox, Nigel de Grey and Frank Birch, there was a fresh new generation and approach coming through. Previously, codebreaking had been the domain of classicists, those who were accustomed to piecing together papyri and other fragments of long-dead languages and bringing them back to life. But the emergence of Enigma had changed a great deal. Here was a machine that posed not a linguistic challenge but a mathematical one. It would need a fresh infusion of mathematicians to tackle the seemingly insoluble problem of breaking through those potential thousands of millions of combinations of letters.

Among the first of the new recruits, some months before war was even declared, was the young Alan Turing, who had already published *On Computable Numbers*, had held dialogues with the philosopher Wittgenstein and had spent time in the US at Princeton. Another prize appointment was that of a young mathematics lecturer at Sidney Sussex College Cambridge, Gordon Welchman, who, unusually for Bletchley Park, cut a rather handsome and dynamic figure. Welchman was at the forefront of scooping up as many suitable maths undergraduates from their universities as possible. (Chess players were also much sought after).

And one such candidate was a young man hailing from Belfast called John Herivel, who had demonstrated a precocious talent for mathematics aged ten. Gordon Welchman had been his lecturer at Sidney Sussex and had spotted his talent. Like other maths whizzes, such as Peter Hilton and David Rees, Herivel found himself transplanted from Cambridge to a small town in the north of Buckinghamshire, which lay on a busy railway junction and which was chiefly notable for the manufacture of bricks. Bletchley Park itself might have had a sort of early-Edwardian charm, but the town which lay beyond its walls was a solid and unremarkable prospect of terraced houses and a small bustling high street. Like all of his fellow recruits, Herivel solemnly signed the Official Secrets Act in an office at the main house, and he was then assigned a place in Hut 6, which lay across the lawn.

In those early days of 1940, faced with those Enigma-encrypted messages, the only tools were pencils and chalk on blackboards. One of the problems Herivel's mathematical skill bent towards was finding a logical short circuit in the Enigma system. He believed there to be some unobserved quirk of the machine and the way that it was used that could help reduce those horrible millions of potential combinations to something more humanly possible to contemplate. For a time, Herivel and his contemporary David Rees were sent to another department called Elmers School to examine the Enigma's plugboard settings, among the machine's many other brilliant features.

The Enigma was a marvel of elegance and utility, with a keyboard like a typewriter with slots at the top to fit in that day's allotted code wheels, and a lamp board that illuminated the encoded letters. Thousands of these machines were being used right the way across the German military, their use guided by a strict timetable of codebooks, so each operator would know which wheels to use on each day. The Enigma both encrypted and decoded. An operator at one end would type in the plain message and the machine would give each encoded letter in turn; these were then radioed over to the distant recipient and the operator at the other end (with his machine set up in the corresponding way) would feed in the encrypted letters and the Enigma would illuminate the letters of the plain message.

Here was a machine that could be used in the desert or in a submarine, at the head of an invading force or in a simple dugout in a forest glade. It was a masterpiece of brass and Bakelite. And the Bletchley codebreakers examined their old models forensically. They assessed the way that the electric turn of the wheels and the rotors would ensure that no single letter would ever be encrypted in the same way. An A turned into a B would not do so again; instead the next time A was used, it might be a Q or a W. This meant that it was no use looking for patterns of the most frequently used letters. And was there anything in the machines, brass contact points that might provide any sort of clue? Herivel, by his own later admission, became madly competitive about the knotty problem. This was not an occupation that one could switch off as soon as the shift came to an end. Herivel and all the other codebreakers knew, rather, that the success or failure of their efforts could determine the fate of the nation.

On a cold and frosty night in February 1940, John Herivel returned to the house in which he had been billeted. In common with all other Bletchley landladies, Herivel's host had absolutely no idea what it was that this young man in civilian attire was doing for the war effort. She might well have known better than to ask, for this was a time in which discretion and secrecy were understood as being necessary. However, her curiosity must have been intense. There had been local rumours that Bletchley Park was in fact a special lunatic asylum for geniuses.

The landlady would certainly have seen how tired her lodger's shifts made him. When he arrived home, she ensured that he had a good hot supper and a pot of tea. Then she left him alone with his thoughts. And that was how he came to be occupying that small front parlour on the quiet frost-crunching night, staring into the burning coals of the fire. He possibly even dozed a little in the waves of heat. But his mind was still churning. And it was then that he saw it – he saw the Enigma machine being set up by its German operator. He saw how the operator selected the wheels and determined the ring setting. This was a three-number sequence (the numbers on metallic rotating rings) that provided an extra layer of security. He saw the operator attach the code wheels and put the lid of the machine down. And he saw the operator – a German youth, possibly a little lazy, or under stressful orders to hurry it up – failing to adjust or change that three-number sequence, thinking that it would be fine as it was. The machine was ineluctably brilliant and unbeatable on its own terms, but those who operated it were flawed and human and more than capable of mistakes. This deeply imaginative vision of how the Enigma was used in the real world could be invaluable, because if the ring settings could be narrowed down into clusters, then new paths into the codes began to open up.

After a sleepless night, Herivel took this idea to his colleagues at Bletchley and there was immediate intense interest. A 'Herivel Square' was drawn up, resembling older cryptological grids with the alphabet running along the horizontal and vertical edges. The difference was that this plotted out the letters of the indicator settings (sent by Morse and intercepted) and marked x where the lines intersected on the grid. Visible clusters were discovered and the mathematical result of this was that the possible combinations of ring settings to check were narrowed from the many thousands to more in the region of twenty or thirty. This meant that each day's coded messages could theoretically be attacked by hand.

At first there was disappointment. The system, when tried on fresh batches of intercepted messages, did not seem to work. But then Germany invaded the Netherlands and the coded messages came in faster than ever and suddenly Herivel's insight was proved correct. Under the eyes of the codebreakers, Enigma 'Red' Luftwaffe traffic was swimming into clear focus. Within days, Hut 6 was breaking huge numbers of messages. It got into such a rhythm that interceptions brought in at dawn could reasonably be expected to be decoded by later that afternoon. The secret heart of the Nazi war machine stood exposed, the scrambled communications of the Luftwaffe snatched from the air and made readable once more.

The technique gained a name: Herivelismus. Gordon Welchman told his former student, 'This will not be forgotten.' And yet for decades it was. Because of official secrecy, Herivel could never tell his dying father what he had done in the war. His father accused him of having achieved 'nothing'. He could not know that his son's achievement was incalculable. What Herivel did was show the codebreakers that they were not mad to imagine that they could take on the impossible and win.

40. HOW POLAND SAVED THE WORLD

Deep amid the dappled, scented glades of the Pyry Forest near Warsaw, Poland, on 26 July 1939, the most precious secret was handed over. Just weeks before the Nazi invasion that would bring the shadow of death over the entire country, Polish cryptologists were preparing to hand over intelligence to their British counterparts. Far from any witnesses or spies, Alfred 'Dilly' Knox and Commander Denniston of Bletchley Park met the men who had achieved the impossible. For it was the Poles who had first unlocked the possibility of cracking every Nazi code.

The mathematicians were Marian Rejewski, Jerzy RòŻycki and Henryk Zygalski. They had been wise enough to keep very quiet about their triumphs. And now here, in this forest, as the distant thunder of war threatened, and with the British pledged to support Poland in the event of that attack, the fruits of their genius were shared. Dilly Knox, a man of volcanic temperament, was intensely displeased to learn that one of the Polish methods used was one that he himself had thought of and then abandoned. They had made it work. The bespectacled Marian Rejewski, a thirty-three-year-old mathematician, had been working for the past seven years upon the Enigma challenge. By the early 1930s, and the rise of Hitler, the system had been adopted by all branches of the German military. Poland – intensely vulnerable as it lay between the aggressive regimes of Germany and Stalin's Russia – had to forge a head start in self-defence.

Rejewski, together with his young colleagues RòŻycki and Zygalski, had been approached by the Polish cipher bureau in 1932 to attempt the impossible. They had to figure out the workings of Enigma without actually having an up-to-date machine to work with. There was some aid from the French secret service to do with the basic functions and nature of the machine, but these were the only clues. It took Rejewski and his associates just a few weeks to imagine their way into the job. They literally envisaged secret wiring systems in their own heads. And amazingly, they did so with real speed. For the layman it was impossible to conceive the scale of what Rejewski was achieving. He was thinking about matters such as 'permutations' and 'cycle structures' and all of this was worked out with chalk on blackboards.

These barnstorming mathematicians delved into what they understood as the machine's one weakness: the way that the introductory indicator code was utilised (in other words, the setting that the machines receiving the codes would use). It consisted of any three letters, say PRV, that might have been selected for that day. Then the operator would send those letters twice (PRVPRV) as confirmation. Enigma never encoded the same letter the same way twice, so the result would look like a convincingly random six letters of gobbledegook. But it also shortened the mathematical odds. Whatever the gobbledegook, the codebreaker would know that the first and the fourth of those six letters were originally the same. This also went for the second and fifth, and the third and the sixth – there were pairs of repeating letters. Suddenly the proposition was not a frightening infinity of millions upon millions of possibilities. It had (however slightly) been narrowed down. The Enigma was not invulnerable.

In their Warsaw workrooms and laboratories, these Polish codebreakers went further still. It took a little more help from the moles in French intelligence supplying them with real-world Enigma indicators and settings for particular days for them to begin to open up the machine's mysteries: for instance, the way that the keys were wired. But with these sequential clues, summoning sense from coded gibberish started to look ever more possible. And by 1937, Rejewski and his team were moved to a secret concrete base built in the Pyry Forest, south of the city, amid the undisturbed trees and quiet woodland roads. In this safe location, they could now start launching their technological fightback against Enigma. Rejewski had already developed a marvel called the cyclometer, which electrically motored its way through permutations. All the success with indicators and keys and the envisaging of Enigma's wiring meant that messages could now be decoded. But there also came forth a new device called the *bomba*. This was in essence an amalgam of six Enigma machines, which given a prompt of probability ran through Enigma keys and unlocked them with ever-increasing frequency.

Zygalski too was contributing to the ingenuity of detecting regular patterns in electric chaos. His invention was Zygalski sheets: special thick sheets of paper upon which a large square was divided into 51 × 51 smaller squares and covered with the letters of the alphabet. Again, this was to do with permutations and the starting positions of the Enigma rotors. Where there was consistency in certain letters, holes would be made in the thick paper, the sheets would then be placed on top of each other and eventually a light would be shone from beneath. Where light was visible through a hole, there would be the key to breaking into rotors and settings.

This was all early days and the Nazi war machine was constantly refining and improving its fleet of Enigma machines, making them ever more secure, meaning the codebreakers also had to evolve their methods. But under the intense summer sun, amid the insects and the chattering birds of the forest, the arrival of the Bletchley team at the small concrete huts was an acknowledgement that all this extraordinary wisdom would now have to be spirited back to Britain as there was no way they could risk an invading Nazi force (which as it turned out was just five weeks away) uncovering just how far these brilliant young Polish mathematicians had thwarted their codes.

And as that war got underway, the lives of the Polish mathematicians were fraught with jeopardy. At first, they were spirited to France, where they continued to plough away at the ciphers (and Rejewski corresponded with Alan Turing). Then, with the Nazi over-running of that country, they were flown to Algeria. This was only a short respite because they were then returned to Vichy France under assumed names (though in the course of all of this frantic movement, RòŻycki was killed on board a ship that had been attacked). This terrifying fugitive life, pretending to be French while still working on codes, was not sustainable. Eventually, there was a desperate bid to cross the Pyrenees to get into Spain. After much jeopardy, including a corrupt guide, they succeeded, but found themselves in a Spanish jail for three months until the Allied authorities at last found them. By 1943, Rejewski and Zygalski were in England.

But there was a horrible twist to their entire meandering war: for curious bureaucratic security reasons, they were not seconded to Bletchley Park. Instead, they were posted to an exiled Polish Army base in Boxmoor. Such codebreaking work as they were given involved working on the much simpler hand ciphers used by the SS. The hideous irony was that it was their ingenuity that had first made the Bletchley triumph possible. It was their *bomba* machines and Zygalski sheets that been taken up and used on an industrial scale. It was their brilliant optimism that had first told Bletchley personnel that the Enigma could be smashed. So it was a tragedy that they were not allowed to take their rightful places at the heart of the codebreaking centre. For a great many years, Rejewski and Zygalski had no idea just how much their ideas and inventions had suffused Bletchley. They had no way of knowing that adapted bombe machines were now being used at bases dotted around the north of London on a factory scale, that they were working on codes that had been intercepted in all corners of the world, or that they had given Bletchley an amazing omnipresence. No thanks came their way.

And worse still, because of the intense official secrecy surrounding the entire cryptology operation, their absolutely invaluable contribution remained deep in the shadows for years afterwards, unspoken and uncelebrated. It was only in the 1970s that Rejewski, who was living in by-then Communist Poland, was able to start telling his story and to start receiving the recognition that had been denied him for much of his life.

It must have been a small consolation. For these Polish mathematicians had changed the course of the war – and of history – before that war had even started.

41. THE MEN WHO ENCODED THE CENTURY

It was in some ways an uncanny coincidence. Many miles apart and total strangers to one another, there were once two men thinking very similar and brilliantly inventive thoughts about the future of cryptography. One was a sparky inventor called Arthur Scherbius, from Frankfurt, Germany, the other a Swedish entrepreneur, Boris Hagelin, who was born in Azerbaijan. Between them – and their intellects – they helped to shape the fortunes of the Second World War. One of them would come to make a very tidy sum for his part in the pageant of history, but the other would die prematurely before seeing the impact that his extraordinary encryption device would have. The key to both of their breakthroughs was the way that rotors in a cipher machine could be made to move independently. And they were both thinking along those lines in the immediate aftermath of the Great War.

These days, the lesser known of the two is Boris Hagelin, creator of several versions of a niftily small cipher machine that was taken up by the US military: the C-36, the C-52 and the M-209 models, which would see later service in the Korean War. His machines had casings that made them look like gun metal sandwich boxes. Inside were tightly bunched rows of rotors, each with letters of the alphabet. This was a device intended explicitly for battle stations. It was compact and durable, and electrically powered rotors and wheels produced an unfathomable number of potential combinations. But the young Hagelin was swept away by the tides of history. He was twenty-five years old when the Russian Revolution of 1917 and its wide repercussions caused his family to leave Baku and return to Sweden. Hagelin's father invested in a new cipher machine firm and he got his engineer son Boris a position there, to make sure that the investment produced returns. Young Hagelin gave intense thought to the principles of cryptography and he built upon designs that the firm had already been working on.

There was a whole range of commercial possibilities, from insurance firms to banks. But Hagelin was also aware of the possibilities for military intelligence. By the early 1920s he had developed a neat little device, the B-211, for the French Army. This, and its later evolutionary models, looked outwardly a little like a typewriter and inwardly like some form of clockwork fantasia. But the interlinking brass components of rotors, letter wheels and cross-wiring were electrical, and there was also a small printer. The paper strip ran adjacent to the keyboard and gave the encoded letters. By the 1930s, as the atmosphere in Europe darkened, the French ordered large quantities of these machines. By 1939, Boris Hagelin found another grateful and sizeable client in the form of the US Army, who were sold on his even newer model, the Hagelin BC-38 and M-209. Resplendent in their khaki cases, these devices were dazzlingly complex and featured twin printer strips: one tape showing the plain text, the other showing the resulting encrypted letters. This was to be the US Army's favoured mode of cryptography across a wide range of theatres of war.

In the new freezing geopolitical climate after that war, Hagelin's continuing inventiveness in a world of ever more sophisticated electronics produced a whole new generation of devices that found intensive espionage use. It might be said that the various iterations of his cipher machines were the secret baseline of so much communication at the time. The wider world knew nothing of these machines, but passing through them were secret messages that affected everyone's lives.

Perhaps the greatest contrast between Boris Hagelin and Arthur Scherbius was that Hagelin succeeded in making a vast amount of money out of his cryptographic advances. Scherbius sadly died too early to enjoy great material gain. However, out of the two, he was the one who subsequently acquired immortality. For while few are familiar with Hagelin machines, everyone knows about the design classic that is Enigma.

Scherbius was an enthusiastic engineer and inventor. Before turning to the mysteries of cryptography, he had developed ceramic heating parts and 'electric pillows', but his real breakthroughs began with his work on what were termed asynchronous motors. This in turn led to a lightbulb moment in devising an electric encryption machine that was simultaneously easy to use and impossible to crack (for the time being!). The trick of it lay in the movement of the rotors and in the electrical wiring that connected them. He added a component that came to be known as a 'reflector', which gave Enigma another unique feature: no letter would ever be encrypted as itself.

In the wake of the First World War, the German Army was dissolved, but in the 1920s it began to reform and rebuild. Within its intelligence ranks were those who understood that the British had managed to break into their codes during the course of the previous conflict and they knew that they would need a new cipher system. Enigma looked ideal. It was portable, useful in any sort of climate or terrain and it generated so many millions upon millions of potential letter combinations that no human intellect on the planet could match it. By the 1930s, it had also been adopted by the Italians and it was being used in the Spanish Civil War. When the Nazis got hold of Enigma, they refined it further with extra flourishes. Yet Arthur Scherbius was not to see any of this. He was killed in a horse and carriage accident in the leafy Berlin suburb of Wannsee in 1929, aged fifty.

Enigma – and the story of the fight to break it – remained deep in the shadows until the early 1980s. But Arthur Scherbius could scarcely have predicted what would have happened as soon as that veil was ripped aside. A bestselling novel by Robert Harris (with Hollywood film adaptation), a slew of books and permanent exhibits in museums such as Bletchley Park were all to come. His Enigma (rather than his electric pillows) became synonymous with all ciphers, and the workings of those beautiful machines – the keyboards, the encrypted letters lighting up and the brass letter wheels – will always haunt the imagination as the very last word in intellectual puzzles.

1

ENIGMATIC

Shapes have replaced the letters of the word ENIGMA. Which shape stands for which letter? That's your challenge: crack the code and read the message.

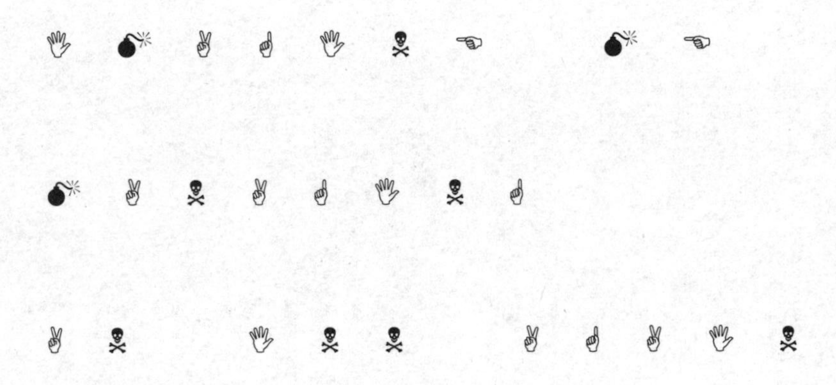

2

IN HIDING

Sometimes a code can be cracked because of what is not there, rather than what is. Look at the sentences below and see if you can crack the code. When you have, you will find that what is missing will give you a clue about a rendezvous.

1 His sister is Agnes, but Arthur's brother is called Fred.
2 The snow thawed as the temperature rose.
3 The performance took place in a small fringe theatre on the edge of town.
4 His character flaws are numerous but at the same time his virtues are many.
5 It is a common misunderstanding but a serious one none the less.
6 She is always impatient and never satisfied with her lot.

3

RECRUITING

In 1942, the *Daily Telegraph*'s challenge to solve its cryptic crossword in twelve minutes or under led directly to successful solvers being recruited to Bletchley. Eighty years on, here's a cryptic for you to solve. There's no time limit, but there *is* a message waiting to be discovered when the grid is completed.

CLUES

ACROSS

7 Cost for military advance (6)
9 No dole confusion, nitwit! (6)
10 Interpreter ran away with lost art (10)
11 Painting, sculpture, writing star turns (4)
12 Shrewd able to before New York (5)
13 Informed and dumped (6,3)
15 In tears, we hear, describing a wedding cake (7)
20 Afterwards friend thinks outside the box (9)
21 Go wrong or make a mistake (5)
23 A piece of building land forming storyline of a play (4)
24 Son finds ten in Angola and unearths ancient Briton (5-5)
25 Statement with regard to gospel writer (6)
26 Trainer has a look back before the Queen (6)

DOWN

1 Thelma right in keeping in the warmth (7)
2 Smoothing out this Board? (7)
3 Dame also appears at lunches for example (5)
4 Just like these clues! (9)
5 Move in direction of hospital rooms (7)
6 Ancient argument! (3,4)
8 Instantly asleep and plunged into darkness (3,4,1,5)
14 Bail Al also known as a music producer (9)
16 Uncontrolled anger followed nobleman before time (7)
17 Ate arty concoction needed for afternoon repast (3,4)
18 Reacted badly having produced something original (7)
19 Soho to Venice exposes where you feel the heat (3,4)
22 Sounds like constraints come from eggs (5)

4

LISTENERS

Coded messages coming through the headphones added an extra skill needed to crack codes, and the Bletchley geniuses could use this skill. Solve the clues below. If you can do that, then you can crack this code. It is not the spelling of the solution which counts, it's the sound it makes.

1 Decomposed vegetable matter used to enrich soil. Pulled tight. Chop a block of wood. Also. Water or marsh plant. Percussion instruments consisting of brass plates.

2 Beef, lamb or pork. In that place. Luxury resort on the French Riviera. Perceive sounds. Be dressed in clothing. Female sheep. From Helsinki.

3 Organ of sight. Young woman. Person who constantly complains. Squeeze moisture out of fabric. Consumed food. Removes the skin from an orange.

4 Sorceress. Part of the plant which is under the ground. Forest. Strong wind. Crushes food with the teeth.

5 One of many you grind to make coffee. A pair. Wander. Having an unfair advantage. Total amount. Herb. Chimney. Farewell. Ordinary.

5

ACCESSORIES

Daphne, Ida, Muriel and Vera are four young women who have been recruited from their universities to join the codebreakers at Bletchley Park. On a summer evening, they decide to unwind and picnic on the lawns. The fashion-conscious females take the opportunity to dress in their finery. They each have gloves, handbag and hat. All the accessories are black, brown, green or white.

All the pairs of gloves are different in colour, all the hats are different in colour and all the handbags are different in colour.

The ladies have the surnames Black, Brown, Green and White. No lady has an accessory with a colour matching their surname.

Each of the Bletchley girls sports accessories in three different colours.

Muriel's surname is the last in alphabetical order, whereas her handbag has a colour that comes first in alphabetical order.

Vera Black sports a green hat and Ida wears brown gloves.

Can you match the ladies with their accessories?

The party will have to break up as one of the ladies is needed for urgent coding work. She is the lady sitting opposite from Ida, and in between the lady with the brown handbag and the lady with green gloves. Who is she?

6

MATHEMATICAL MINDS

Back in the days of the First World War, the art of codebreaking was by and large seen as the territory of classically trained scholars. The emphasis was on knowledge of language and the ability to decipher codes seeped in ancient history. By the time of the Second World War, the emphasis had changed. The mathematical masterminds had moved centre stage.

In this addition sum, numbers have been replaced by letters. The letters are taken at random; there is no particular significance in the choice of letter. Each individual letter stays constant in numerical value. So for example, if U were equal to 9, every U in the sum would represent a value of 9.

Can you crack the code and work out what number each letter stands for?

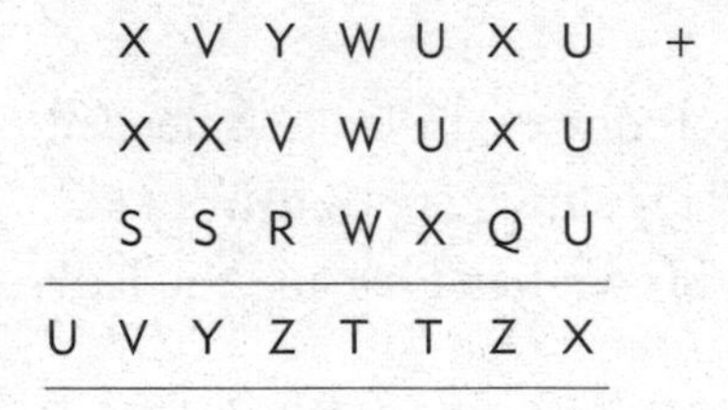

7

TYPING POOL SLIP-UP

The following message has just come in from the typing pool. It doesn't make any sense as it stands and looks like just a jumble of numbers. However, there is a typist's logic here; it's just a case of working out what it is by working out where the typist has slipped up.

23 4317843 6974 17346 59 03534 5603 85 975

8

CROSSING THE RIVER

The following four puzzles were devised by 'Alice in Wonderland' author (and mathematician) Lewis Carroll, who was often cited by Bletchley codebreakers as an intellectual inspiration.

Four gentleman and their wives wanted to cross the river in a boat that would not hold more than two at a time.

The conditions were that no gentleman must leave his wife on the bank unless with only women or by herself, and also that someone must always bring the boat back.

How did they do it?

This is probably not an original problem invented by Lewis Carroll, but it is clear that he liked logical puzzles of this kind and often used them to entertain his friends. Another favourite was the famous river-crossing problem involving the fox, the goose and the bag of corn, which he often gave to child-friends to solve. The problem of the gentlemen and their wives was given to colleagues at Oxford.

9

THE FLOWER RIDDLE

'This time [Alice] came upon a large flower-bed, with a border of daisies, and a willow-tree growing in the middle.

'"O Tiger-lily!" said Alice, addressing herself to one that was waving gracefully about in the wind. "I *wish* you could talk!"

'"We *can* talk," said the Tiger-lily, "when there's anybody worth talking to."'

Three particular friends of Lewis Carroll were Harriett, Mary and Georgina Watson, the daughters of the Reverend George William Watson. A number of puzzles and games were invented for this trio, including this poetic riddle:

Tell me truly, Maidens three,
Where can all these wonders be?
Where tooth of lion, eye of ox,
And foot of cat and tail of fox,
With ear of mouse and tongue of hound
And beard of goat, together bound
With hair of Maiden, strew the ground.

Can you identify these wonders?

10

THE SUN AND THE MOON

Another 'Puzzle from Wonderland', which was published in *Aunt Judy's Magazine*, concerned a dialogue between the sun and the moon. Again, the riddle is written in verse:

Said the Moon to the Sun,
  'Is the daylight begun?'
Said the Sun to the Moon,
  'Not a minute too soon.'

'You're a Full Moon,' said he.
  She replied with a frown,
'Well! I never *did* see
  So uncivil a clown!'

(Query. Why was the moon so angry?)

11

CATS AND RATS

This problem came from an article which Lewis Carroll contributed to a magazine called *The Monthly Packet* in February 1880. The article begins with a problem which has four possible solutions depending on the interpretation of the original information. In all four cases, the consideration of fractional cats and rats is ruled out. The problem is as follows:

> If 6 cats kill 6 rats in 6 minutes, how many will be needed to kill 100 rats in 50 minutes?

Identify the four different ways in which the cats might kill the rats, and determine how many cats will be required in each scenario.

CHAPTER TWELVE

THE STAR-SPANGLED CRYPTOGRAPHERS

In which the extraordinary ingenuity of American codebreakers proved an inspiration to friends and allies, from the Battle of Midway to macabre medieval manuscripts.

42. PURPLE HAZE

There was horror in paradise. The skies had darkened, and the attack had come like lightning. In December 1941, molten flames and billowing black smoke rose high above the rich blue waters of Hawaii. The American naval base at Pearl Harbor had been turned into a glowing inferno by the Japanese Air Force. This was a moment upon which the course of the world's conflict was hinged, and the ripples that continued on after this great wave hit would prise open the Japanese military campaign and render it naked and vulnerable.

As the British (with the brilliant help of three Polish mathematicians) were burrowing into Germany's Enigma codes, the Americans were unlocking the mysteries of Japanese encryption. Earlier in 1941, even as America maintained its neutrality in the war that was unfolding, a substantial team of cryptanalysts were gathered at a classified subterranean operations room at Pearl Harbor. As well as mathematicians, there were musicians. Just like the British, the Americans had discovered the curious relationship between codebreaking skill and aptitude for musical composition. The Japanese military had adopted a code system known as JN-25 (with an evolved version called JN-25B) in 1939.

One of the chief difficulties was that unlike Enigma – where the British had at least managed to secure a couple of the machines, and thus investigate its myriad possibilities – the American codebreakers had to work with snatched photographs of the Japanese 'Purple' and 'Magic' (as they were termed) enciphering machines. Levering these ciphers open was a matter of small, incremental successes. But there was also technology to hand – the US codebreakers had brilliantly adapted some of the new state-of-the-art IBM tabulation machines to help sort through a range of coded possibilities. This was just as well because no human memory could have coped with the vast amount of cross-referencing of Japanese military terms and vocabulary that had to be painstakingly stitched together to form a logical picture.

And there was also help from the Australians and the British, among others. The Australians naturally had a direct interest because Japan could very easily have turned its attentions upon that continent. Meanwhile, at Bletchley Park, Colonel John Tiltman, of whom we will hear more elsewhere (see page 286), and Alan Turing were also trying to reason their way into the Japanese codes. They realised they were dealing with a five-digit cipher but could not penetrate any further. When the Japanese attack on Pearl Harbor came, the US codebreakers under Captain Joe Rochefort could not have anticipated it from the tiny fragments of messages that they had so far managed to decode.

The attack, which claimed over two thousand lives and sent mighty battleships to the bottom of the sea, not only finally brought America into the war, but also redoubled its secret efforts too. By early 1942, simply through painstaking inference and pinpoint-sharp cross-referencing of every single deciphered term, the US codebreakers had pulled off the remarkable feat of uncovering about 20 per cent of the Japanese naval codes. The secrets were beginning to yield to them and success rose exponentially. Now it was time for the American Navy to pull off the most amazing coup in deception and double bluff.

In the middle of the vast blue Pacific was a US military base on a tiny island just several miles wide called Midway. The location was strategically vital for patrolling and policing thousands of miles of sea. Thanks to the intense remoteness of the location, Midway was also nakedly vulnerable. But Captain Joe Rochefort and naval commanders saw the opportunity to draw the encroaching Japanese Navy into a trap. There was an underwater communications cable that was completely untappable running thousands of miles between the Midway military base and Hawaii. Using this cable, the US Navy, together with the codebreakers, instructed Midway to send out a radio message that wasn't jumbled into code but put out in plain English. This message was akin to a distress call, claiming that the base's desalination plant had broken down and, as a result, Midway was facing an imminent freshwater shortage. This message went out on the general airwaves and the codebreakers were able to detect that the Japanese had picked up on it by their own messages discussing the base and using the terms such as 'water' and 'desalinate'. The base was too tempting a target for the Japanese Navy to ignore. To take it out meant that they would have uncontrolled mastery of the Pacific and supply routes to America and Australia.

As well as the IBM computers and a highly talented team, Captain Rochefort also had extensive experience of Japanese language and culture. One of the marvellous lateral means of unlocking a code was due to the Japanese tendency towards extreme politeness and deference. Messages to superiors would begin with highly elaborate courtesies, such as 'I have the honour to inform your excellency . . .' Once deciphered, they gave the codebreakers further clues to the keys that were being used. (And in a pleasing irony, this was precisely the opposite of the clues offered to the British by the Wehrmacht. The codebreakers realised that the young Enigma operators, when putting in the new code settings each day, would send test messages to other operators employing hearty German obscenities. As Keith Batey observed, 'Yobs will always be yobs,' And so anyone who made a study of German swear words would be able to use them as a crib.)

The Japanese Navy made its plans for the attack. It was to come on 4 June 1942. And when that array of battleships converged on Midway, the US Navy was ready in force. The battle as it unfolded was terrible, but this time for the Japanese forces. There were horrendous casualties on both sides, but broadly speaking, it was Japanese planes and shipping that were sent plunging into the dark Pacific depths. Aircraft carriers became towers of flame. On one, Admiral Nagumo was said almost to be in a trance or a fugue state, reluctant to leave the ship even as fire roared through its innards. The ferocity of the fighting extended over two days. American fighters were atomised as well, but it was the once-superior Japanese force that in the end was forced into retreat.

The humiliation was so intense that the defeat was initially kept secret from the Japanese people. Only the Emperor Hirohito and the highest echelons of command were informed about the full scale of the disaster. It was the turning point of the war in the Pacific, and one of the advantages that the codebreakers had gained for the ever-expanding US forces was that Midway Atoll became an even more useful strategic base. Submarines sailing out from Hawaii could now use it as a refuelling point, giving them a much wider range.

And while the battle itself was an extraordinary example of naval forces fighting with extraordinary courage thousands of miles from substantial land, out under a measureless sky and on ruthless waves, it was also a landmark moment for the US cryptographers. It was an example of how a combination of diligence, lateral thinking and ingenuity conducted in secret basements could produce profound changes in the outside world.

43. THE UNCANNY MANUSCRIPT

The images have something of the alchemist's laboratory about them. Human figures in weird-shaped jars, plants and flowers that defy identification, unsettling images of nude people gathered in strange glass chambers, and women bathing in curious green liquids. This cavalcade of surrealism is accompanied by a close, small handwritten text, across many yellowed vellum pages, that looks just on the edge of familiarity and yet, on closer inspection, resembles no other language. This is a book – or codex – that was said to have been pored over by candlelight at the sixteenth-century court of Emperor Rudolf II, amid the snows of Prague. Running to some 240 pages, and measuring 20 centimetres by 16 centimetres, its re-emergence into the light of the twentieth century ensnared and bewitched the finest minds in American (and British) codebreaking. Nazi ciphers were one thing, but this eerie gothic artefact was quite another. And the Voynich Manuscript has continued to exert immense fascination and influence because it would appear to be a medieval code that has outwitted all subsequent generations.

The Voynich Manuscript is currently housed in the Rare Books Library at Yale University in the US. Its vellum is without question centuries old, but that seems to be the only certainty around this extraordinary book. Apparently lost for centuries, it was discovered and acquired by a Lithuanian dealer at the turn of the century. His name was Wilfred Voynich and he was a communist who had fled the pogrom-haunted wastes of nineteenth-century eastern Europe to set up book dealerships in London and subsequently in New York. Voynich was said to be on a visit to Rome when he happened across the manuscript in an old seminary. The illustrations and the writing were both vivid and weird, made up of exquisite, unidentifiable flowers, swirling patterns, tiny human figures in what looked like test tubes. There was more still . . . a letter was tucked within. Written by one Johannes Marcus Marci and dating from the mid 1600s, it claimed that the manuscript had formerly belonged to the Emperor Rudolf and thence to an alchemist called Georg Baresch. This was an extra dimension of verisimilitude. In circumstances that remain hazy, Voynich acquired the manuscript and eventually took it back to New York.

And it was not many years after that that the Voynich Manuscript was brought to the attention of William Friedman, one of America's foremost experts in codebreaking, whose contribution to the later war against Japan would be beyond value. Friedman was long accustomed to facing alien languages and dialects with mysterious alphabets. The manuscript was a challenge that he could not turn away from. As a result, it was to perplex and beguile both him and his codebreaking wife, Elizabeth, over the course of several decades.

Legends had begun springing up about the book. There were some who claimed that it was the work of the thirteenth-century philosopher and proto-scientist Roger Bacon. There were others who thought that its provenance was later and yet darker, and that instead it was an occult document from the hand of Elizabeth I's 'sorcerer' Dr Dee (see page 123), an effort on his part to penetrate into the forbidden mysteries of spirits and angels. And there emerged, a few years later, even wilder rumours. One suggested that the manuscript had been written by something that was not of this earth. That it was in fact an alien artefact bringing news to humanity of alien life, and alien language and culture.

The weird inscrutability of the document wholly seized the imagination of Friedman, who, by the time of the Second World War, was one of the world's foremost cryptological thinkers. And as that war ended and the codebreakers of the US and the UK turned to face the chilly new climate of the Cold War with the Soviet Union, Friedman formed an after-hours casual codebreaking club with many of his senior associates, some of whom had returned to academia. Photocopies were made of the Voynich Manuscript. Efforts were made to transcribe every single character of every single word so as to begin the most rudimentary frequency analysis to see if it was possible to spot encrypted 'and's and 'the's and commonly used vowels. There were discussion groups conducted after long days of intelligence cryptanalysis. What made the problem so attractive was the aesthetically pleasing nature of the manuscript. Not just the naked illustrations but also the formal beauty of whatever language it was written in.

In 1951, Friedman reached out across the Atlantic to one of his most trusted Bletchley counterparts, Brigadier John Tiltman. By this time, the wartime organisation of Bletchley had now evolved into GCHQ (see page 328). Brigadier Tiltman, who sometimes favoured bright tartan trousers as part of his regimental uniform, was also an extraordinary cryptologist. At Bletchley, it was he who had worked out how the Japanese ciphers might be broken simply by using the power of his imagination. He had also devised a means by which new young recruits could be taught enough functional Japanese in just six months to be able to start cracking codes themselves.

So who better to start applying themselves to this mad treasure? The Friedmans and their informal club by this stage had arranged the unknowable 'letters' into tables, but no patterns were immediately discernible. Tiltman was immediately hooked. He thought the weird illustrations of plants and fronds that were threaded throughout the manuscript might yield a way in and took a copy of these images to a senior horticulturalist, who not only could not identify them but also became rather agitated about the fact that he could not do so. Yet a couple of the images did suggest familiarity, such as the one that looked a bit like bindweed and another that may have been a sunflower. Though as Tiltman was to point out, this would have given the manuscript quite a late date, as that flower was not introduced into Europe until the late fifteenth century. But what of the manuscript's zodiacal and astronomical illustrations? In the 1920s, it was being suggested that one such starry image depicted the spiral nebula. This would have been quite a feat for the early medieval eye to see before the development of telescopes.

Tiltman was also engrossed in the history of the manuscript itself. He discovered a fragment about Queen Elizabeth's court astrologer and philosopher Dr Dee examining it. A quote from Dee's son Arthur, conveyed by Thomas Browne, described how, when in Bohemia, he pored over 'a book . . . containing nothing but hieroglyphicks; which book his father bestowed much time upon, but I could not hear that he could make it out.' Had the queen's dark necromancer himself been pulled into the unceasing whirlpool of fascination that the manuscript generated?

There was another possibility that both Tiltman and Friedman explored: was the manuscript an early demonstration of a new kind of 'universal' language, like the twentieth century's Esperanto? In his thrillingly diligent research, Brigadier Tiltman came across a book written in the 1600s called *The Universal Character*, by a man called Cave Beck, one of the founding members of the Royal Society (and therefore dedicated to the pursuit of knowledge and science). Beck declared that he was just the latest in a line of men who wished to see a language developed that could be understood the world over. Was it possible that such a language could in part be formed by numbers?

But this still did not quite explain the mysterious words of the Voynich Manuscript, for it seemed impossible to identify any form of regularity about them. There were other theories being advanced even as Tiltman studied it. One suggested that the words on the pages might be some form of ninth-century Cornish, and someone else suggested that they were part of an old Aztec tongue. Another idea was that these words were survivors of an older Proto-Romance language. All of which was very well, but still no one was any closer to solving the mystery.

Tiltman was intrigued by the weird plant images and explored that particular frond of history. There were parchments and papyri dating back hundreds of years. In AD 63, a horticulturalist called Krateuas served Eupator, the King of Pontus, and he began a tradition of 'herbals': these were manuscripts that detailed plants, flowers and herbs and explored how they might be used medicinally. Subsequent 'herbals' that appeared throughout the Roman Empire became more lavish. The illustrations, sometimes in vivid colour, had their own unearthly beauty. Attention was paid to roots such as mandrake, but their depiction was stylised and plants could sometimes be given human features. There were also astrological symbols interspersed that told of when, according to the stars in the heavens, certain herbs would be at their most effective. This might have gone a long way towards explaining the Voynich illustrations. If they could be understood as allegory, then perhaps their meanings might begin to be discovered. But even then, they proved stubborn.

The latter years of the twentieth century brought more searching technology into play. The Voynich vellum was subjected to radiocarbon dating. The result showed that the manuscript dated from about 1404 to 1438. That would have ruled out as a potential author Roger Bacon, who had lived around one hundred years previously. And it would also have ruled out Dr Dee, who was not born until the following century. One magnificently devious possibility was that it was a hoax perpetrated by Wilfred Voynich himself. The immediate objection to this is that if the vellum was provably from the fifteenth century, how precisely could he have done that without the aid of a time machine? The answer is that he would not have needed one! All that would have had to happen was for Voynich to come across a supply of antique blank vellum. From this point, having acquired the aged material, he could then begin using suitable ink materials (which would fade and appear to age in time) to cover it with beguiling nonsense-writing and images.

It is a lovely idea, but quite aside from the monumental effort involved to produce such a magnificent hoax work of post-modern art, what would have been in it for Voynich? As a rare book dealer, there were easier and more straightforward ways of making money. On top of this, the vision encapsulated in the manuscript, even down to the depiction of naked figures, seemed determinedly that of another time. It would have been tricky to fake this convincingly on such an epic 240-page scale.

And so the quest goes on. Quite recently, a German expert in Egyptology, Rainer Hannig, delved deep into the Voynich enigma and came up with a new and intriguing possibility that the manuscript was perfectly genuine, and that the curious language was in fact based upon Hebrew. 'Countless decipherment attempts were made,' he wrote. 'A lot of languages were proposed, such as Latin, Czech or among others Nuahtal (a form of Aztec), just to name a few . . . The word-structure leaves only one possible explanation: the manuscript was not composed in an Indo-European language.' However, the imagery in the book did seem to be broadly European, therefore the language had to have some geographical proximity. It could have been a form of Aramaic or Hebrew, both of which, he declared, were spoken by European scholars, and the structure of those coded sentences were also more suited to a Semitic language.

This fresh possibility set off a new rumpus of codebreaking. Aramaic and Hebrew scholars examined the Voynich Manuscript anew. Its significance as a code lies in the fact that across the years the world's finest cryptologists have come to regard it as the Everest of encrypted conundrums; a challenge too tempting and too beautiful to resist.

A good fifteen years after first clapping eyes on the Voynich Manuscript, Brigadier Tiltman was giving illustrated lectures to highly select groups of codebreakers, discussing the possibilities and how delightfully elusive they were. Many of those who shared his enthusiasm in that hermetic world knew that his was the mind that could prise open an encryption through sheer force of visualisation. He had an amazing and almost unmatched talent for seeing the jumbled chaos in front of him and arranging it into fresh, regimented systems that would at last bring order, sense and meaning to it. He could even visualise the workings of encryption machines that he had never before seen.

But here was a problem that with a certain grace continued to dance away from him. It danced away from William Friedman too. Despite all recent breakthroughs, there is every chance that it will nimbly elude giving up any of its meanings. Voynich's manuscript changed the world of codebreaking by demonstrating that there could be such a thing as an unbreakable cipher that was in its own terms a hauntingly beautiful work of art.

44. THE DOTS AND THE DASHES

Even Dr Dee would have been thrilled by the words materialising from the very air itself. In 1838, the American crowd that gathered around a small factory in New Jersey were gripped. The phrase 'a patient waiter is no loser' was not in itself memorable or profound. But the means by which it was transmitted that day changed the entire course of communication. Here was a new code that within a few years would sweep the world and alter the course of economics and of warfare. Presiding over it was a man who had previously been an extremely talented oil painter: Samuel Morse. He would give his name to a code that became a much-loved language.

Sending messages down wires through the means of electrical impulses, known as telegraphy, was the technology that the great generals of antiquity had dreamed of. To be able to hold a conversation across a distance of many miles, instantaneously, without the need for messengers and horses, was almost magic made real. Samuel Morse had first had the idea while on a long ocean voyage back to the States from Europe. A fellow passenger, Charles Thomas Jackson, was experimenting with electromagnetism and had assembled a makeshift laboratory on board for various experiments. The fascinated Morse saw the possibility of using this power to send messages along a single wire.

Ideas are sometimes like contagions. There can be outbreaks of identical notions occurring in different countries and in different continents at roughly the same time. This was the case with the concept of electric telegraphy, and Morse found himself up against German and English competition. As he worked at the problem of sending messages long distance, from city to city, others were developing their own systems. In 1837 in England, Messrs Cooke and Wheatstone combined their shiny new miracle technology with another modern marvel – their electrical telegraph was used on the new London to Birmingham railway. But Morse's system, which swiftly evolved so that it was possible to send messages from Baltimore to Washington DC (guaranteed to grab the attention of government), had an elegance and an ease of use that other systems lacked.

The code in its earlier form involved spools of ticker tape, an inky stylus and a delicate tick-tocking mechanism. The idea was that the telegraph would be sent and received by this automatic ink-and-tape apparatus. There were lines and dots and the way that they were indented was an indicator of the characters that they stood for. The machine itself was one of those nineteenth-century things of beauty, made of brass, wood and clockwork. Soon it was operating on railways and across great swathes of the country. But Morse's colleagues looked at it and could see ways that the system might be made even more speedy. In particular, the clicking and the ticking of the receiving apparatus could be fashioned into a code that might be transcribed directly by the listener. It was from here that the distinct character of Morse – written down as dots and dashes representing each letter and transmitted as 'dits' and 'dahs' – evolved.

In later years, the energetic Morse would claim in grinding court cases that the chief elements of electrical telegraphy were fundamentally his and should always be treated as his copyright or patent. Born in 1791, Samuel Morse seemed in some ways emblematic of the new independent American. He had an all-consuming work ethic that was sometimes channelled in the most unexpected directions (his career as a portraitist was no less impressive; in fact some of his paintings were adjudged masterpieces). With Morse telegraphy came further advances that were previously unimaginable: underwater cables running along the seabed right the way across the Atlantic. And from there emerged a whole new branch of espionage, tapping such cables could be 'tapped' thus apprehending the Morse messages. The military applications were vast and varied. Use of Morse reached its peak in the Second World War, making some quite extraordinary demands of those who were required to learn it, to send it and to translate it.

As the young British women who volunteered for the navy and the army in the Second World War found, Morse required a very great deal of time and attention. Those who were drafted into the 'Y-service' (Y was short for 'wireless') and who would spend the war intercepting messages in Morse which they would transcribe and send on to Bletchley Park, first faced a gruelling training course. Recruits were chosen for their mental agility, and were sent to a training camp on the Isle of Man, where they spent months learning Morse code. They became so immersed in it that it would form part of the wiring of their brains. Morse operators in the field were sending out their encrypted messages at high speeds, and those who intercepted them had to work just as quickly; there could be no hesitation over a single letter of the transmission.

Furthermore, the Morse messages were pre-encrypted and so it was almost like a code on top of another code. Whereas in straightforward Morse messages, the meaning would begin to unfold on the page as the listener instantly translated the dots and the dashes into letters and words that ran in the right order, wartime Morse messages were scrambled and jumbled so that each individual letter was completely random and bore no relation to the last. In that sense, the operators were translating and transcribing gibberish. Nonetheless they had to do so with the purest accuracy, for the slightest slip could mean the difference between life and death.

There was a young woman from north London called Pat Sinclair who had a youthful enthusiasm for electrical apparatus (back in the 1930s, this was not an attribute that was encouraged among young ladies). As such, she was swift to volunteer as a Secret Listener. The role enabled her to do her wartime bit effectively and would also feed her fascination for the new technology. She could also clearly see how it might lead to a huge amount of stress for any particular interceptor if care was not taken. Pat and her fellow listeners were put to long hours hunched over headphones, their wrists aching from writing letters in pencil, and their minds and eyes straining to keep the focus completely on the coded letters coming over the ether. This was a role that required young brains, though there were some older Morse interceptors too. Many succumbed to burnout and nervous complaints. Yet at the same time, there was the reward of knowing that what they were doing was making a very tangible daily difference to the war effort. And on top of this, the secret listening bases and outposts frequently had youthful atmospheres that lent themselves to off-duty laughter and romance. (One of the unusual attributes of Pat Sinclair's listening base – HMS *Flowerdown*, in Hampshire – was that young Women's Royal Navy volunteers sat alongside male interceptors. They worked as pairs, so that they could cover all intercepts more thoroughly. But this also opened out opportunities for weekend dances and dalliances).

There were other Y stations right the way across Britain. From the underground base at Scarborough on the Yorkshire coast, to makeshift huts on the south coast, Secret Listeners were tuning into enemy aircraft and sometimes receiving romantic messages from pilots, who knew that young women would be listening to them! Morse itself, though composed of binary noises, could also be used to slyly convey emotion simply by means of its transmission: each and every operator would have a distinctive style of sending out messages, and those listening in would learn to distinguish between these different operators. This distinctive operation style was referred to as a 'fist'. The German operators always knew that their radio Morse messages were being intercepted. Curious abstract relationships formed over the dits and the dahs, young German operators smiling at the prospect of the young women listening in, and the young women in turn envisaging the Morse senders, whose 'fists' they could always identify.

It was a long war and the effort of the Secret Listeners was intense. There was an unexpected after-effect in the years following the conflict. Pat Sinclair recalled that Morse code had so seeped into her brain that it remained second nature. The result was that whenever a film or TV drama had a sequence of Morse messages being sent, she could instantly translate them – sometimes faster than the characters on the screen! There were other occasions too when she could tell that the producers had simply made the messages pure gibberish.

In a cyber age, it ought to be the case that Morse code is now a museum piece, and yet there are still many enthusiasts out there, as proficient as ever in working with this ingenious form of encryption. It changed the world because, for a time, Morse code was part of the soundscape of that world. Those staccato dits and dahs bore witness to terrible battles and mighty victories, to ruthless invasions and brilliant counter-attacks.

45. THE WHISPERS ON THE WIND

Sometimes secret codes can be the poignant echoes of languages that have been suppressed by conquering powers. There are words and phrases and cadences that have been smothered by oppressors and, as such, can be used by their original speakers in the spirit of defiance. This was the case for the languages spoken by the many tribes of the Native Americans in the early years of the twentieth century, whose lives and cultures had been decimated by the swarming colonialists taking over the whole of North America. The new rulers of the west decreed that the children of Native Americans should be turned away from the old tongues. They were enrolled in boarding schools far from their own homes to be educated out of old family ways. Yet somehow the languages survived. And out of this cruelly implacable conquest came a rather beautiful cryptography development in the Second World War.

Languages such as Navajo and Comanche were suddenly called into fresh life during the war years when they were needed to communicate secrets. These were tongues that the Japanese would not have the first idea about. The languages were used in substitution ciphers, one word standing for a particular letter. And in faster moving theatres of war, the languages would be spoken as they were when they were formed – a swift and brilliant means of conveying messages and intelligence. Sometimes the poetry of these crushed languages even found fresh flowering. The men who became the 'code talkers' found themselves at the heart of the conflict and their contribution was invaluable.

The idea had been around since the Great War. And indeed, in the inter-war years, as the Nazis ascended to power in Germany, they too took an interest in the possibility that the languages of the Native Americans offered (there is a suggestion that Goebbels considered these peoples 'Aryan'). A few German linguists were assigned the task of immersing themselves in Comanche, both to get a feel for it and also to see what code possibilities presented themselves. Some, posing as graduate students, came specifically to the States to try to discreetly find ways of studying it. Yet with bitter irony, it turned out that there was little written material for them to draw from. The nineteenth- and twentieth-century erasure of these native languages was so thorough that there was no literature to be found on them. The Comanche people had, among themselves, managed to keep the traditions and the language alive, but there was little for the outside world to see.

A further bitter irony was that the American government had good reason to be grateful that these and other native tongues had not been so thoroughly exterminated as intended. And fortunately, some of the young Native Americans who had been through those obligatory government boarding schools to turn them away from their old traditions were perfectly enthusiastic about throwing their skills into the conflict that was coming. Some of the traditions that the US authorities had sought to stamp out were warrior rites of passage. The young Comanches and other Native Americans had never forgotten them and here was an opportunity to offer their courage on their own terms, with skills that only they possessed.

When America was finally pulled into the war in December 1941, following the Japanese bombing of Pearl Harbor, the code talkers were swiftly assembled. Indeed, some had been brought together a year earlier. As an example, a company of two dozen Choctaw speakers from Oklahoma had been mustered and they had been put through the rigorous training of signals intelligence. Together they worked to contrive a brand-new code that would employ the Choctaw language, while at the same time keeping the meanings opaque. Because of new modern technology, there were some English terms that had no equivalent in Choctaw – and the result was an outbreak of delightful (and cryptic) similes. For instance, they would make references to 'sewing machine guns', and mighty 'turtles', all under the direction of 'Crazy White Man'. Even if enemy interceptors picked up these terms unencoded and managed to translate them from the Choctaw, what could they have made of them in any event? 'Turtles' referred to tanks and 'sewing machine guns' were in fact machine guns. And Hitler would have been displeased to hear that he was the 'Crazy White Man'.

The brilliance of the code talkers was to reach its height in the summer of 1944, and in the bitter final battles beyond as the Nazis were finally vanquished. Teams of Choctaw speakers, Comanche speakers and Hopi speakers were among the thousands who, with grace and courage, crossed from England to France on 6

June 1944 – D-Day. At the start of the desperate battle for the soul of Europe, the code talkers proved brilliantly agile and dextrous in getting vital intelligence over to units moving across all sorts of terrain. With radios, headphones and field transmitters, these signals operatives were on those deadly beaches as the landing parties were attacked, moving through countryside infested with a still venomous enemy, and as they spread out through France, Luxembourg and Belgium, their exquisite languages formed an indispensable web of unbreakable communication. The hazards all these men faced were akin to those that the agents of the Special Operations Executive were also challenged with. There were gun battles, mortar attacks and the constant threat of injure, capture or death. Even so, through all of this, the code talkers provided incredibly fast, on-the-ground intelligence. The commanding officers in several platoons were profoundly impressed with their lateral-thinking recruits.

As with all matters cryptanalytical, the tragedy of the story of the code talkers is that the brilliant work they did, and the courageous feats they pulled off, had to remain a secret for many years afterwards. Who knew if such secret codes might be needed again? Therefore many of the code talkers had to wait a great many years before they got the full public recognition and the praise that they deserved. There were post-war instances where Comanche men, now demobilised and back in the US, took part in traditional celebration dances where the cloud of war could be ceremonially dispelled from their heads. Happily, their story could finally be told and even now there are branches of the US military that periodically honour this brave fleet of code-carriers. After all the bitter years of violence and suppression in the nineteenth and twentieth centuries, as Native Americans were harried and dispossessed, this at least is a small recognition of the richness of the culture that the United

States had once sought to wipe from the land.

46. THE WOMEN WHO BROKE STALIN

Paranoia and fear insinuated their way through the Moscow streets like a contagion. Throughout the city at night, frightened men and women held their breath in the dark at hearing unfamiliar footsteps in the communal corridors outside. During the 1930s, this was a city of disappearances. People from every walk of life were taken from their homes in the small hours to face terrifying ordeals of interrogation, torture and frequently death. In Stalin's Russia, no one was sure of the ground beneath their feet. That was partly because their absolute ruler was himself in the grip of paranoia. When it came to the matter of secret ciphers, it was later to transpire that he had good reason to be.

By the time war came, a particular encryption system termed 'one-time pads' was in use by the Soviet secret service, the NKVD (and would also be used by its successor, the KGB). It was complex but did not require complex technology. It was based on the principle of one-off codes, printed on separate sheets, with a serial number at the top of each page for reference. The idea was that a message coded using truly random keys used only once could never be broken by enemy agents. The agents out in the field would use these encryption sheets in a pre-agreed order. Transmuting letters into numbers, they worked from the (sometimes very small) sheet to turn their messages into chaos. This meant that no possible pattern or repetition could be discerned and there was no giveaway of commonly used vowels or terms. If used properly, then it was wholly unbreakable.

And there was also an element of old-fashioned secret-agenting in the ways that spies would get rid of the one-time encryption sheet once they'd encrypted or decrypted a message. One method was burning, and another was swallowing. Soviet agents (and indeed British SOE operatives who employed a similar system under the guidance of code genius Leo Marks) found ways of carrying these one-time pad code keys within innocuous books. This was a system that needed only a pencil and paper to work and so discretion was relatively easy. It was suggested that some one-time pads were so small that they could be concealed within the shells of nuts (a very strong magnifying glasses would be needed).

So throughout the war, Soviet agents operating from Portugal to China relied on this fantastically effective system out in the field. And because from 1941 they were Allies of the western powers, there were noises made about the different friendly nations refraining from cracking one another's codes. But there were those in the US who felt deeply that the Communist Soviets would not be friends for very long. And even as their partnership in the war against Nazi Germany got underway, small teams of US codebreakers were picking over Soviet coded transmissions with care.

Within the halls of the Kremlin, Joseph Stalin seemed sceptical and distrustful of the codebreaking efforts of others. When fresh intelligence on Nazi manoeuvres and plans in Russia were presented to him, he initially appeared reluctant to believe it. Yet it was his own agents' codes that became acutely vulnerable from 1941, as the Nazi war machine smashed and burned its way east towards Moscow. The remorseless advance that left villages ablaze and countless thousands upon thousands coldly massacred put an enormous amount of pressure on all areas of the war effort, and one of those was the secure production of the one-time pads. At the time they were treated a little like printed banknotes. Produced in conditions of security, the specially printed pads and their random sequences of numbers (plus the all-important serial numbers at the tops of the pages) were dispatched, via couriers and embassies, to all the corners of the earth. But under the pressure of that nightmarish Nazi advance, the one-time pads stopped being printed in pairs (for sender and recipient) and instead a number of them were repeated with multiple printings.

Astonishing though it may seem, once the Americans and the British had detected through covertly acquired pads that many were being used again, it presented a chance. In 1943, it was the American codebreakers who could not resist diving into it. They would have argued that it was vital to understand both Soviet movements and Soviet thinking. Operation Venona began, and it would continue for the next few decades. These codes would go on to have seismic after-effects, causing spies and traitors to be unmasked and governments to be rocked. All this despite the fact that their unravelling remained the deepest darkest secret.

At the centre of the genesis of Venona was a twenty-three-year-old schoolteacher from Rose Hill, Virginia. As soon as the US entered the war, Gene Grabreel, who had a keen mathematical mind, was eager to do what she could. Thanks to a young army officer whom she knew and bumped into at the post office, she had been swiftly selected to be sent to Washington for induction and training. The secrecy was deep and absolute. All her father was allowed to know was that she had been recruited to 'shuffle some papers around'. In fact, she ended up at an establishment called Arlington Hall in Virginia. The US equivalent of Bletchley Park, the Hall had once been a girls' school that had now been requisitioned. As a result, many of its young female former schoolteacher recruits looked as though they might still be teaching. Gene Grabreel had come from a quiet, respectable home where chickens were kept and now she was in the epicentre of the secret war. Training was very fast. She and a great number of her fellow recruits had a mathematical feel for the discipline that could only be improved by the experience of attacking real codes.

There were large rooms with bays of recruits. Most were working on Japanese ciphers. However, Arlington Hall's Carter Clarke, the head of army intelligence, grimly calculated the need to read the minds of the Soviets, and so around Gene Gabreel there soon formed a team of brilliant women who seemed to thrive on what to most would be a nightmarish challenge. One-time pad material was being obtained from all sorts of obscure corners. For a long time, it really did appear as though the Soviet system was completely invulnerable. That was until the work of another terrific recruit, Mary Jo Dunning, who ran some machine tests on a commandeered IBM punch-card system, detected that the messages, although impossible to unscramble, showed signs of what was known as 'depth' or glimmers of a recognisable pattern. The machine demonstrated, in fact, that some of the one-time pads had been used more than once.

What then ensued was an extraordinary collective act of logical deduction combined with lateral leaps of thought. Among the women of Arlington Hall, there were some brilliant linguists who could lend the weight of their expertise with Russian, and others who had been studying the provenance of thousands of coded messages sent to and from the US, using the one-time pad system. The IBM machine had indicated that there was a golden prize that was worth pursuing, but it could not by itself put all the pieces of the enigma together. That took the living minds of young men and women who, when not focused on the tiniest scraps of encoded messages, were relaxing by playing bridge or going for walks in the woods. All of these young people, when seen together, simply looked like mature students.

Arlington Hall became an extraordinary trove of intelligence and the teams who worked on the Venona Project (as this codebreaking operation came to be known) achieved great feats of cryptological ingenuity. By gathering in messages from all conceivable locations and from a huge variety of agents, they began to peer into the darkness of a world so secret that they did not even tell the US president about it. Security was so paramount that the Commander-in-Chief himself could not be trusted not to accidentally give their breakthroughs away to the Soviets. Venona uncovered enemy spies, including Ethel and Julius Rosenberg – US citizens who had passed vital atomic secrets to Soviet handlers in the 1940s (for which they were subsequently executed) – and Klaus Fuchs, a scientist who had worked on the Manhattan Project to create the world's first atomic weaponry (the bombs which, in 1945, were dropped over Nagasaki and Hiroshima). Fuchs had memorised vital elements in research and had also passed this intelligence over to the Soviets. In both cases, their capture was down to a confluence of intelligence, but the Venona decrypts were kept extremely quiet.

It was also thanks to Venona that treachery at the heart of the British establishment was revealed. Decoded messages from spies revealed that there were double agents at the centre of UK intelligence. More clues built up the mosaic and soon it revealed that these traitors were Guy Burgess, Donald Maclean and Kim Philby – who would later be bracketed together as the 'Cambridge Spies'. The revelations were horrifying and practically no one in authority was allowed to know the exact source or provenance of this intelligence. There were two or three people in Britain's codebreaking community who were privy to Venona, but that was about it.

And that secrecy paid off because the codebreakers were mining Venona encryptions right up until the early 1980s. They had spent almost forty years obtaining the most amazingly close-up intelligence from behind the Iron Curtain, not only without any Soviet leaders knowing but without most western leaders knowing too. In that sense, Venona was also a great triumph for that conspiracists' favourite: the deep state. A new world of electronics brought new cipher systems and the one-time pads came to seem prehistoric in the face of computers. But fascinatingly, even then, those messages remained worth mining. The Venona operation was one that changed the course of America and Britain's secret worlds, unleashing decipherments whose effects seeped out over years.

1

OVAL OFFICE

The Oval Office at the White House is the working space for the President of the USA. This code substitutes letters for other letters by using a fixed number of places moved in the alphabet. The first group stands for OVAL. Can you shape up and work out all these words, which have a common link?

1 U B G R

2 I O X I R K

3 Y W A G X K

4 U H R U T M

5 X N U S H A Y

6 Z X O G T M R K

7 V K T Z G M U T

2

CAMBRIDGE SPY SQUARE

One of the most notorious scandals of the Cold War concerned the Cambridge spy ring. Members of the ring had all been students at the University of Cambridge. In 1951, two of the group, Guy Burgess (codename Hicks) and Donald Maclean (codename Orphan), fled England bound for the Soviet Union.

Solve the seven clues below, writing your answers vertically in the grid. When the grid is complete, the letters in the middle row will spell out the name of the ship on which they left Southampton on the midnight sailing, bound for northern France on the first leg of their journey.

1 Stumped, unable to progress further with a problem or conundrum.
2 Betrayal of one's native land.
3 One of a number of university buildings on campus.
4 Country, capital Beirut, from which spy Kim Philby fled to Moscow.
5 First name of Donald Maclean's American wife.
6 Official language of the former Soviet Union.
7 Government Office for which Maclean had worked as a diplomat.

1	2	3	4	5	6	7

3

CODER

Each letter of the alphabet is represented by a number from 1 to 26. We give you the numbers which represent the letters in USA to start you off. So every space with a 1 in it contains a U, every space with a 2 contains an S, and every space with a 3 contains an A. The completed grid will be made of words about the USA that are written either across or down and interlink as in a conventional crossword. There are some well-known place names included

The checklist below will help to keep track of the letters you have found.

1 = U, 2 = S, 3 = A, 4 = , 5 = , 6 = , 7 = ,

8 = , 9 = , 10 = , 11 = , 12 = , 13 = , 14 = ,

15 = , 16 = , 17 = , 18 = , 19 = , 20 = ,

21 = , 22 = , 23 = , 24, = , 25 = , 26 = .

When you have filled the crossword and worked out all the letters, look at the numbers below and find the name of a place linked to code-cracking.

3. 25. 21. 16. 9. 12. 13. 24. 9. 20. 3. 21. 21.

■	26	■	2	■	20	■	■	■	15	■	■	■	22	■
20	3	15	3	16	16	■	6	4	3	21	■	19	1	9
■	2	■	12	■	21	■	3	■	8	■	21	■	4	■
13	3	7	16	■	21	3	2	2	24	■	3	25	4	3
■	14	■	9	■	2	■	2	■	■	■	11	■	9	■
13	4	7	3	2	■	8	3	25	2	■	4	3	2	13
■	9	■	15	■	17	■	8	■	16	■	2	■	■	4
8	3	18	■	15	16	8	20	16	13	3	■	1	2	3
1	■	■	18	■	4	■	1	■	4	■	10	■	15	■
18	4	3	25	■	15	4	2	13	■	5	16	17	4	2
■	14	■	24	■	■	■	4	■	24	■	26	■	4	■
2	16	12	9	■	13	24	13	4	6	■	8	16	13	23
■	2	■	7	■	25	■	13	■	3	■	24	■	2	■
20	24	13	■	8	24	26	2	■	20	1	14	2	24	9
■	9	■	■	■	23	■	■	■	3	■	4	■	26	■

4

EAST TO WEST

The names of six US states are hidden in the sentences below. Can you plot the coded journey beginning in the most eastern hidden state and ending with the state in this puzzle which is furthest west and south? We have mixed the states up so the journey is not in the order of the sentences, and not all the states on the journey are mentioned.

1 They are following in Diana's footsteps, working hard for others.

2 I first met Norma in England in the 1950s.

3 Rules made law are rules which must be strictly adhered to.

4 She exclaimed, 'How splendid a house is that?'

5 Can Eva dare to take a flight in a hot air balloon as she hasn't a head for heights.

6 He knows the territories in the Balkans as well as his own homeland.

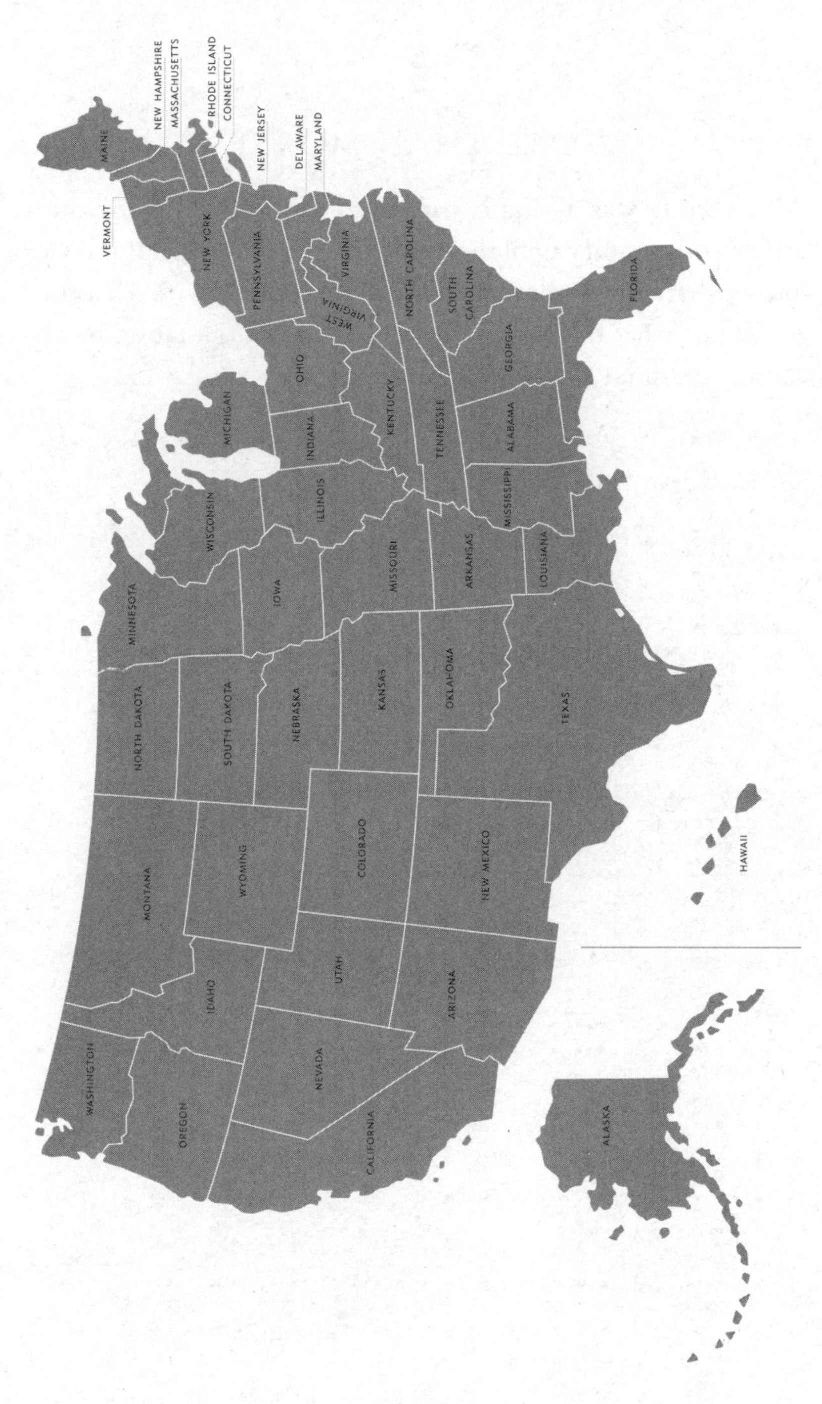
MAINE
NEW HAMPSHIRE
MASSACHUSETTS
RHODE ISLAND
CONNECTICUT
VERMONT
NEW YORK
NEW JERSEY
DELAWARE
MARYLAND
PENNSYLVANIA
VIRGINIA
WEST VIRGINIA
NORTH CAROLINA
SOUTH CAROLINA
FLORIDA
GEORGIA
OHIO
MICHIGAN
INDIANA
KENTUCKY
TENNESSEE
ALABAMA
MISSISSIPPI
ILLINOIS
WISCONSIN
MISSOURI
ARKANSAS
LOUISIANA
IOWA
MINNESOTA
NORTH DAKOTA
SOUTH DAKOTA
NEBRASKA
KANSAS
OKLAHOMA
TEXAS
MONTANA
WYOMING
COLORADO
NEW MEXICO
HAWAII
IDAHO
UTAH
ARIZONA
WASHINGTON
OREGON
NEVADA
CALIFORNIA
ALASKA

5

MORE THAN MORSE

Morse code was a vital communication tool from the mid nineteenth century until the end of the twentieth. In this code puzzle there is a bit more than Morse to contend with. Can you work out what this is and reveal a five-word quotation by an American about loyalty and patriotism?

▽▽ ▽△▽▽

▽△▽△ ▽▽▽ △△▽ ▽△ ▽ △▽△ ▽△▽▽

△▽△ △△ ▽▽△ △△△△ ▽

▽▽▽ △▽△

△▽▽ △▽△ ▽▽▽ ▽△ ▽▽△

MORSE SYMBOLS

Alphabet

A •–	G ––•	M ––	S •••	Y –•––
B –•••	H ••••	N –•	T –	Z ––••
C –•–•	I ••	O –––	U ••–	
D –••	J •–––	P •––•	V •••–	
E •	K –•–	Q ––•–	W •––	
F ••–•	L •–••	R •–•	X –••–	

Numerals

1 •––––	3 •••––	5 •••••	7 ––•••	9 ––––•
2 ••–––	4 ••••–	6 –••••	8 –––••	0 –––––

6

MAN OF LETTERS

Solve the quick clues and transfer the letters to their correct slots in the grids. You can then read a quotation by a famous US author and the shaded squares can be rearranged to form his name. His first name has four letters and his surname contains five letters.

1	2	3		4	5	6			
7	8	9	10		11	12			
13	14	15		16	17	18		19	20
21	22	23		24	25	26	27	'	28
29	30	31	32		33	34	35	36	
37	38	39		40	41	42			
43	44	45	46	47	.				

CLUES

Wed, not single 1.2.31.45.44.10.9
Less strong 4.16.5.27.35.36.46.18
Ate well at celebration 20.15.8.6.12.26.47
Reflections, but not in the sun 28.14.11.39.19.24.42
Express gratitude 21.22.41.3.32
Male pig 34.30.37
Departed 40.25.17.43
Cut grass 7.38.33
Moist, damp 29.23.13

7

THE STRAIGHT AND NARROW

Crack this code to find some capitals of US states by figuring out the capital letters represented by the code below. It's the shape of the letter which counts – its curves and its straight lines. S stands for straight line, C stands for curve. Not as easy as it looks as some letters have the same combination of curves and straight lines. Which letters will keep you on the straight and narrow?

A – S3 (it has three straight lines), B = C2S1 (it has two curves and one straight line), C = C1 (it has one curve), D = C1S1, E = S4, F = S3, G = C1S1, H = S3, I = S1, J = C1S1, K = S3, L = S2, M = S4, N = S3, O = C1, P = C1S1, Q = C1S1, R = C1S2, S = C1, T = S2, U = C1, V = S2, W = S4, X = S2, Y = S3, Z = S3.

a) C2S1 S3 S2 C1 S3 C1S2 C1 C1 C1S1 S4 (two words)

b) C2S1 C1 C1 S2 C1 S3

c) C2S1 S1 C1 S4 S3 C1S2 C1 S3

d) C1S1 S1 S4 C1S2 C1S2 S4

e) C2S1 C1 S1 C1 S4

f) S3 S3 C1S2 C1S2 S1 C1 C2S1 C1 C1S2 C1S1

CHAPTER THIRTEEN

CODES FROM THE FUTURE

In which we trace the secret codes that led to the first computer and analyse mysterious ciphers from outer space. And finally, a look at the challenges that face today's cryptographers, and the codes that lie deep within us all.

47. THE COMPUTERS THAT BEAT THE FÜHRER

The borough of Poplar in east London was a noisy, foggy and pungent prospect at the beginning of the twentieth century. It lay close to the busy docks of the River Thames and its tributary the River Lea, with water that sometimes ran orange with industrial pollution and weed-trailing shores that housed clanging metalworks and vast gasworks. The air rarely cleared, and all around the muddy streets and market it was dense and flavoured with smoke. This was the environment that produced a man who in his efforts to break the codes of the Nazi regime created a machine that would lead the way towards a cleaner, less industrial world.

It was also this unpromising cockney background that led to the man's colleagues not believing that he could be a genius, for surely geniuses went to Oxford and Cambridge, rather than studying at night school? This Poplar man was instrumental in bringing the computer into being. It was all part of his mission to crack the technologically complex Nazi codes that were used for messages sent by Hitler and his top lieutenants. Had Britain had a little more national foresight, perhaps this man's work would have headed up a turbo-charged world-beating effort to develop computer technology after the war. As it was, his achievement was kept in the shadows of secrecy for many decades.

That young man was Tommy Flowers. He was an engineer with a particular gift for thinking his way around practical problems. After early promise and the grinding hours he put in to both his work and his education, he found himself taken on by the General Post Office, which at that time also oversaw the evolving technology of telephone and other electric communications. A national network of wires and exchanges posed its own series of logistical challenges, and it was in Tommy Flowers' time that it first became possible to make direct calls, rather than routing a call through a human operator. He and his colleagues occupied some modern, light and airy laboratories in a specially built institute on top of a hill in north-west London. Thanks to such matters as the tapping of telephone cables, this GPO research unit in Dollis Hill was conversant with the demands of national security. And come the war, it was engineering centres such as this that would be called upon to help codebreakers build fantastical new code-checking machines. In 1940, a firm in Bedfordshire specialising in office tabulating equipment oversaw the super-secret development of the bombe machines for Alan Turing and Gordon Welchman.

But later in the war, a new type of Nazi code was brought into play, one that at first glance seemed even more impenetrable than Enigma. These were the Lorenz ciphers. Electrically generated and working on a binary principle, the machine that would be needed to unscramble them would have to be constructed to a similar level of complexity. A young firebrand mathematician called Bill Tutte and an older, wiser professor, Max Newman, began thinking their way into the conundrum, with contributions from Alan Turing, who had already envisaged a world of computers. What would this programmable beast look like, how would it work and how reliable would it be? It fell to Tommy Flowers, as an expert and sharply inventive engineer, to take the concept on board and give it physical form.

His capacity for working all hours came into its own and what gradually emerged from his intensely practical imagination in 1943 was a machine that was dubbed the 'Heath Robinson' because, with its maze of ticker tape, spools and flashing lights, it resembled one of the mad-scientist creations of that cartoonist. In theory it worked, but in practice the results were a little bumpy. Although the machine could process the code data it was fed, it was temperamental. It relied on two reels of paper tape running absolutely simultaneously, yet the tape was vulnerable over the course of hours to stretching. Nonetheless, when fed encryptions from the German Lorenz code-generating machine, the apparatus did produce results. Prior to this, there were those at Bletchley who had rather prematurely dismissed Tommy Flowers on the grounds that 'the Crafty Cockney', as he was snobbishly termed behind his back, couldn't possibly have the intellectual power for the task. But Flowers persisted, and he knew that he was right.

The idea evolved, and by 1944, it had become Colossus. In essence, this was one of the first programmable computers and it was a revolution of machinery. The fact that it was capable of chewing through binary Lorenz ciphers and decoding messages sent from Hitler's desk almost as soon as they had been intercepted was one thing. But it also represented the future. If this was possible, what might follow? And the Colossus worked because Tommy Flowers had the genius stroke of incorporating extra valve technology, basing it upon one spool of tape as opposed to two, meaning the system could be kept running continuously without juddering faults or halts. He also built the whole thing by hand. The 1,500 or so valves gave it a quite extraordinary speed compared to the Heath Robinson. Here was a machine that might now run through the night, for endless nights, being fed codes and producing the decrypts. (And for some of the Wrens tending the machine throughout those nocturnal shifts, it found an extra use – thanks to the heat it gave off – for drying underwear.)

The contribution Colossus machines (with their further-developed Colossus II siblings) made in the final acts of the war was immense. These were the machines that were reading messages sent by Nazi High Command at unfathomable speeds just as the D-Day invasion unfolded. Indeed, the day before that extraordinary campaign began , on 5 June 1944, Colossus decoded a message from Hitler which revealed that he did not want to move German troops to Normandy as he was convinced the Allied landings would take place elsewhere. From that point onwards, the eleven Colossus machines got so close into the encrypted messages from Hitler and his lieutenants that it was sometimes almost as if the British codebreakers had been present in the meetings alongside them, listening to every word. This continued right the way through to the end of the war in 1945, with the Colossus machines registering the last spluttering flames of Nazi resistance as the Red Army closed in on Berlin.

And all this while, the man who had built these extraordinary machines continued with his work on valve technology and other electronic developments back at the research laboratory in Dollis Hill. At the end of the war, there was a whisper within the hermetically sealed community of codebreakers that all the Colossus machines had been destroyed for various obscure security reasons. But this was not quite true. Certainly nine of them had been dismantled and their parts reused, but two survived and would be moved into the regenerated Bletchley Park – that is, GCHQ in Cheltenham – in the early 1950s. They were used until 1959, when they were superseded by even more dazzling codebreaking computer apparatus.

But there must have been more than a twinge of sorrow within the heart of their creator Tommy Flowers. In the months after the war, he was in debt because the scepticism in the Bletchley hierarchy had forced him to build Colossus using his own money for many of the parts. He eventually received a payment of £1,000 from the government, but it was said that this still did not quite cover the costs that he had incurred. Worse yet was the obligation of continued secrecy. What he had achieved could never be spoken of aloud. So he continued with his GPO work high on Dollis Hill and his innovations continued. He was partly responsible for the 1950s computer, nicknamed 'Ernie', that randomly selected winning Premium Bond numbers. (In the late 1970s, the band Madness devoted an entire song to Ernie.)

Recognition, when it did come in the 1980s, brought an avalanche of appreciation. It had taken over three decades for the vast achievements of this brilliant electrical engineer to be acknowledged. Tommy Flowers died in 1998 but his name lives on in all sorts of places, from a department at Google to educational units in his home borough of Tower Hamlets, to a large wall-sized mural of his face in the Poplar neighbourhood where he was born. The codes that Tommy Flowers' machines broke not only influenced the currents of military history but also shaped the future of technology. Imprints of his work can be seen in the phones and computers that dominate all our lives today. They were, in part, made possible by the persistence of his vision.

48. THE ENIGMA OF THE GENOME

The human body is composed of millions of codes. The elements within human cells that govern everything from hair colour to height are dazzlingly complex, and the structure of our genes continues to baffle us and excite scientists around the world. The very discovery and definition of the structure of DNA (the double helix) in the 1950s opened up the most extraordinary prospects in medicine. All the body's genes, known collectively as the genome, contain the blueprint of the body. But could scientists find a way of decoding the genome, thus defining and delineating it? This might have been a completely different undertaking to unravelling military messages, but in some curious ways, the principles of this quest were similar. The very idea of decrypting the 'instruction manual' for a human being would in previous generations have been literally inconceivable. But now it is the case that a decoded genome sequence can be happily pinged from computer to computer.

The Human Genome Project, which started in 1990, was an even more epic undertaking than the quest to crack Enigma. For the implications of unravelling the secrets of DNA would be the most astounding advances in medicine. Even now, some years after the great triumph of the Human Genome Project, it is a miracle that is somehow taken for granted by a great many of us. And it had its roots in the nineteenth century. A Swiss scientist called Friedrich Miescher was trying to make a study of white blood cells. To this end, he had commandeered (look away now if you have just had breakfast) used bandages stained with pus and blood. What he succeeded in isolating from these revolting articles was a molecule that he called nuclein. It was in fact what would later be termed DNA, though it would be many decades before this discovery was refined and explored further. In the meantime, the idea of heredity – that of characteristics being passed down from generation to generation – had been scientifically explored in the 1860s by Gregor Mendel. But the question remained: by what mysterious means were these genetic traits transmitted from parents to children?

Rather in the manner of DNA, the research was passed on and evolved through the years. By the 1940s, Erwin Chargaff had fixed on DNA as the means by which all this occurred, Barbara McClintock had been awarded the Nobel Prize for her work on identifying 'the jumping gene' that could move on chromosomes, and in 1951, the X-ray crystallography expert Rosalind Franklin managed to capture, by means of X-ray diffraction photography, actual images of DNA. It was from this point that, famously, Messrs Watson and Crick were able to assemble their 3-D model of the double helix in 1953. But the efforts to burrow deeper into the secrets of DNA soon became analogous to the challenge of codebreaking. The human genome was there to be 'deciphered'. The international Human Genome Project was intent on uncovering the sequences of all the bases of DNA molecules and exploring the maze of links deep within those nanoworlds where the traits of humanity were stored and filed.

And it was not long before this epic deciphering had mapped the thousands upon thousands of genes and the billions of 'base pairs' of chemicals that make up the human body. Unravelled, it was possible to store the information of the human genome on a CD-ROM. In 2001, some were being given away as free gifts with a highbrow magazine called *Prospect*. By 2003, the project was complete. As an achievement, it was almost beyond computation. To have broken the code and then laid out the very blueprint for humanity itself was an incredible step in modern science. It was noted by one scientist that decoding the human genome had opened a door that had never before been seen through and what he could see through it was a number of other secret doors which also required opening! Breaking the cipher of DNA was and continues to be very decisively one of the decryptions that fundamentally affected the course of history.

The medical possibilities that have been opened up are extraordinary. It has given us the chance to isolate and to beat certain forms of cancer that had always defeated previous generations. It also offers the chance for us to explore the complexities of metabolism, and to treat inherited conditions such as certain heart diseases. As documented in lurid newspaper headlines, the genome code has also permitted scientists to create such curiosities as mice that glow in the dark. In some of the coverage, there is an element of the Prometheus myth, for just as Prometheus stole fire from the gods to give to mankind (paying a heavy price), so the deciphering of DNA has given man powers over his own destiny that might evoke, it is suggested, some form of retribution from Mother Nature.

But what is also fascinating is the compulsion to see the essence of humanity itself as a form of code – and once the code is cracked, then a whole range of medical and social ills might be eradicated. The genetic code will loom ever larger in our lives the further the scientists delve and try to open all those other 'secret doors'. Unlike many other human developments, this decipherment provides a rare opportunity for genuine optimism.

49. IS THERE ANYBODY OUT THERE?

There is something breath-catching and haunting about this journey beyond the stars. Two space probes, gliding at 35,000 miles per hour through the jet-blackness of space, are passing by moons and planets, borne and buffeted by solar winds. They will move further and further into the solar system and are destined to continue for centuries more, even possibly still journeying after mankind itself has become extinct. Throughout their journey so far they have been sending back data and images from their travels, including extraordinary impressions of Venus and Saturn. But the Voyager probes need power, and the instrumentation will not last much longer. What will last, however, are the messages that Voyager 1 contains in the event of it ever encountering alien life. Some of them take the form of code, but then to any alien civilisation, all language and recordings and indeed images would be a form of code.

Voyager 1 was launched in 1977 and this intergalactic greeting took the form of a golden record – a form that could survive 40,000–50,000 years on the outer edges of the solar system. 'This is a present from a small, distant world,' ran a message from the then US President Jimmy Carter, 'a token of our sounds, our science, our images, our music, our thoughts and our feelings. We are attempting to survive our time so we may live into yours.'

The daunting prospect of assembling cultural and scientific entries for this record fell to the eminent scientist Carl Sagan and a dedicated team. There were the sounds of humanity, collections of messages spoken in fifty-five different languages and also a greeting from a child (Sagan's then six-year-old son). There were recordings of waves crashing, of the wind, of birdsong and of whalesong. There were pictures of vast numbers of animal species and of different global cultures. There was music that spanned from Middle Eastern folksong to Bach. And there was Morse code, as well as a recording of brainwaves, which themselves resembled a code. Here was an intended encapsulation of Planet Earth: its life, its loves.

It was a shade more poetic than Professor Sagan's previous coded message to the stars in 1974, which took the form of a radio transmission. It was aimed at globular star cluster M13 and was more an exercise in celebrating the remodelling of the Arecibo Telescope in Puerto Rico. The message itself sounded like a high-pitched whistle, varying minutely in its note. But if correctly decoded, this auditory binary code would translate into block images of the double helix, a representation of the solar system showing where the message came from, the atomic numbers of hydrogen, carbon and oxygen, and an image of a human stick figure. The star cluster that the transmission was pointed to was so dense as to suggest that if life was to be found, it might well be there. The setback was that star cluster M13 is 25,000 light years away. So none of us will be around to hear any reply.

The compulsion to wave to our alien neighbours with coded signals is a centuries-old tradition. In the early 1800s, an Austrian astronomer called Joseph von Littrow had the romantic idea of digging vast circular trenches, miles and miles in diameter, in the Sahara Desert. These trenches would then be filled with oil and set ablaze. The fiery circles would be an indication to those looking down from above that humanity wished to communicate. In the twentieth century, others had their own coded messages to send. In the chilly depths of the Cold War in the early 1960s, what became known simply as the Morse Message was transmitted from a new 'planetary radar' array in Soviet Crimea. This code was directed at the planet Venus. When unscrambled, the three-letter message consisted simply of 'Mir' (roughly meaning both 'peace' and 'world'), 'Lenin' and 'USSR'. Presumably the Venusians were supposed to infer from this that Earth was unified under a Communist world government.

In recent years, the messages we have transmitted out across the stars in the form of different codes have become rather more sophisticated. In 1999, 'Cosmic Call' was beamed out from a vast observatory in Ukraine. The basis of this code was mathematics, on the philosophical grounds that any extra-terrestrials who had the technology to hear the signal would themselves also be steeped in mathematics. Maths is a form of universal language. The code itself came in the form of what is termed a 'bitmap' that, once unravelled, would show the building blocks of mathematics, such as pi, radius and multiplication. There were also the fundamentals of physics, biology and astronomy. The overall picture would have suggested that Earth had a highly competent scientific civilisation.

A particular distant planet called Gliese 581c was selected to receive a digital radio signal called 'A Message from Earth' in 2008. It comprised a code that when unravelled would be a form of digital time capsule. Perhaps slightly more startling for the extra-terrestrials was the digital message from NASA sent the same year that came in the form of a recording of The Beatles' song 'Across the Universe'. The song pulsed out into the inky blackness of deep space, towards an impossibly distant star called Polaris. Although this was in part to celebrate the fiftieth anniversary of NASA (and the fortieth anniversary of the song), the idea brought with it a curious romantic tingle. You can almost imagine the alien life forms staring at each other as they listen to this composition from another world.

More music was sent out in 2018 under the banner 'Sonar Calling'. It was beamed at Luyten b, a vast world that is light years from earth and that has been calculated to very possibly being conducive to life. The music sent comprised specially composed pieces from artists such as Jean-Michel Jarre and Nina Kravis. The idea, said the organisers, was a 'celebration, an artistic and scientific experiment and a collective rumination on what it means to be human, and alien'. But this of course throws open an even more fundamental possibility: what if the aliens are in turn sending us their own coded messages to try to make contact? It is an idea that has echoed around popular culture with communication via musical notes in *Close Encounters of the Third Kind* (1977) and Tom Baker's Doctor Who receiving creepy coded messages from ancient gods on Mars, as two examples. This latter story was (in part) inspired by a real-life hair-raising mystery at the turn of the twentieth century involving the genius inventor Nikola Tesla.

Tesla made the claim that with his new-fangled radio equipment he had received a message directly from Mars. In 1901, it was still quite respectable to imagine that there was life on the red planet. 'What man who has ever lived would not envy Tesla at that moment!' wrote one newspaper reporter, imagining the moment when those ethereal pulses came through. Tesla was a firm believer that there was a form of civilisation to be found on Mars, and his quest to communicate with the planet occupied him over the years, in between his fantastical scientific innovations with electricity and radio waves and robots and remote control. The radio genius Marconi was in agreement with him concerning the provenance of these weird pulses from space. Tesla believed that the message was mathematical and that when decoded, it simply ran: 'One, two, three, four . . .' It was later understood that what he had heard were the rhythmic radio waves emitted by interstellar gas clouds, and that the wide galaxy is alive with such noisy phenomena. But even though Mars fast receded as a possibility for first contact, that hasn't stopped more recent scientists listening to code-like messages arriving from vast distances out of that infinite dark.

For some time now, the SETI (Search for Extra-Terrestrial Intelligence) Institute in California – a collective of astronomers, astrophysicists and other scientists – has been working towards identifying messages from other stars and planets (Professor Carl Sagan was one of its pioneers). As well as a broader philosophical interest in the origins of life both on and away from Earth, the institute has run programmes for years monitoring radio and light signals reaching through space, in the hope that some day it won't be the steady pulse from a quasar but instead a specifically directed code that, unlocked, will mark the first bridgehead between humanity and other species. Rather than changing history, this will be a code that changes the entire destiny of mankind when it comes. Keep listening to the skies!

50. WHAT WILL THE CODEBREAKERS DO NEXT?

Sometimes it was submarines prowling the frozen depths of northern seas that intercepted the signals, sometimes there were brilliantly sophisticated listening devices hidden in bushes and aimed at the outside walls of embassies. And occasionally such devices were thwarted by the embassy staff holding their meetings not in normal rooms but within special glass cubes. After the Second World War and the triumphs of Bletchley Park, the need for codebreaking remained urgent and pressing. There were all-new geopolitical fault lines and the Cold War between America and the Soviet Union kept the entire world on the edge of lethal conflict. And with operations at Bletchley Park winding down, the codebreaking directorate had to address the computerised future and how they would combat new threats to national security. Their spiritual descendants are still hard at it today and to an extent have emerged from the shadows.

Legend has it that one of the reasons that GCHQ moved from the London suburbs to Cheltenham in the early 1950s was that some of its senior personnel were terrifically keen on horse racing. The truth was rather more to do with the anxieties of the atomic age. In the event of a nuclear strike, London would be the top target, whereas a town in Gloucestershire would not be directly hit. The codebreakers are still there today, in a modern and beautiful doughnut-shaped base. There are also other GCHQ nerve centres dotted around the country, from Birmingham to Manchester to Bude. The ciphers that the extraordinary minds at work there are pitted against are quite different from those faced by Alan Turing and his colleagues. But the principles remain broadly the same.

More than ever before, we live in an encrypted world. From phones to TVs, internet bank accounts and much else besides, we navigate passwords and secure encryption in order to carry out the simplest shopping transactions. Imagine, then, what this means for the current generation of codebreakers. In the 1950s, the focus of COMINT (Communications Intelligence) and ELINT (Electronic Intelligence) was on nations across the world aligned with the Soviet Union, and whether their encoded messages were gathered via eavesdropping ships and submarines or indeed, a little later, via satellites, there was still a sense that they always knew where to listen. In addition, GCHQ and its American counterpart the National Security Agency had a secret head start on many other countries, some of which were still using old Enigma or Hegelin systems. Because of the all-enveloping secrecy around these departments, no one in the wider world could have known that their most secure encryptions had been prised open.

The codebreakers received lavish investment, and something of the 1950s and 1960s space-age techno-glamour could be perceived in the fitting of Comet and Nimrod planes with the latest equipment to vacuum up secret signals from on high. There was another factor that gave the cryptologists a more favourable audience within the highest echelons of government. Britain's HUMINT (human intelligence garnered by spies on the ground) had been shaken by the treachery of the double agents Philby, Burgess and Maclean in MI6. Human intelligence even without this disaster was frequently patchy. Spies by their natures were contrary souls. By contrast, the intelligence provided from intercepted and decoded messages had a sort of crystalline purity. As GCHQ expert Richard Aldrich pointed out, prime ministers came to love GCHQ because it gave them a feed to intelligence that no one else could possibly know about.

Aldrich was also salty on the nature of the special relationship between Britain and the US. While there was assuredly a lot of mutual respect among codebreakers themselves (in particular those such as Brigadier John Tiltman, see page 286), the two codebreaking superpowers sometimes fell out. The US – being by far the richer partner – had pulled ahead with all its efforts in technological terms. And it needed to lean on the remnants of the British Empire, made up of listening stations from Hong Kong to Africa and India, in order to continue intercepting intelligence from the different corners of the world. However, even amid these tensions, GCHQ enjoyed a level of freedom that made it unique as a government department. With the need to keep the veil of secrecy drawn over the very idea of message interception and cipher-cracking (even though every other country was at it), the department escaped the close and sometimes unfriendly scrutiny from politicians and journalists that other agencies had to tolerate.

And how jealously that secrecy was guarded! Up until the 1980s, it was forbidden for newspapers to even make reference to the initials 'GCHQ'. It only became more widely known after Prime Minister Margaret Thatcher tried to ban trades unions there in 1984. Nonetheless, this was still the same organisation that had seen Dilly Knox solving codes while lying in a bath off a Whitehall corridor in 1915. After Bletchley, a number of the wartime faces – from Nigel de Grey to Joan Clarke – had stayed on. Indeed, Alan Turing's former fiancée had proved a formidable codebreaker and stayed with GCHQ until 1977. And though the targets may have been different, the core disciplines of cryptography somehow remained the same, even in an increasingly computerised age. It still needed intense mathematical talent. It still needed all-enveloping linguistic skills. And it still needed an ability to think laterally around problems in a variety of different dimensions.

With the internet age seemed to come an understanding that anonymous secrecy was increasingly absurd, and with the gradual emergence into public view of the story of Bletchley Park, the powers at GCHQ came to realise that it was preposterous to persist in the shadows. Instead, they now understood that by drawing on that rich and wonderful history, they could persuade the public and politicians alike that their work was just as necessary and in many ways the same as it was then. They were here to protect the security of the realm. And even though the details of what they did naturally had to remain under very heavy wraps – it still remains vital that enemies don't realise that their messages and communications have been compromised – the broader brushstrokes could in fact be a focus for pride. These days, working for GCHQ is no longer (just) a matter of being recruited silently and invisibly. You can go to their website and find out how to apply for various careers there. Despite this, the core requirements remain much as they were when the codebreakers of Room 40 assembled just over 100 years ago. The organisation still requires the very best and most flexible linguists, and the knottiest of mathematical minds.

The codebreaking propositions of the future seem daunting. While the era of computing has brought forth cohorts of computer experts who can apply the most abstract mathematical theorems to cryptology propositions, a new generation of infinitely faster technology is approaching at speed. Quantum computing and increasingly sophisticated artificial intelligence will be deployed as means of making intelligence indecipherable. Must then the codebreakers of the future be robots too?

The answer at the moment would surely be no. After all, the advent of Colossus in 1944 did not instantly make platoons of human cryptologists redundant. Until we truly arrive at a point where computers can pass the Turing Test – that is, fool human interlocuters into thinking that they are conversing with a human and not a machine – then there will always be a need for minds that can think upside down and sideways. In addition to this, organisations such as GCHQ now have an entire private as well as national realm of ciphers to deal with, for example, vast numbers of WhatsApp and smartphone messages, as well as the more official governmental codes.

Threats to the defence of the realm now appear to come from a great many different angles. And to burrow into often intensely personal cryptology systems still requires old-fashioned spycraft, as well as a degree of nerdiness. For some time, it has been not just nation states that the codebreakers have to confront but small and elusive terrorist cells as well. Part of the intense difficulty of their work is not just the codebreaking but anticipating where the next threat might be coming from and focusing on cracking the codes of as yet undeclared foes.

The amount of material that GCHQ analyses now might have thrown the Bletchley codebreakers into a tailspin of despair for a short while. But then the Bletchley crew would perhaps have reminded themselves that they too were cracking codes from every continent, every hour of every day, that the intelligence they accrued filled index cards that in turn filled vast huts, and that no matter the complexity of the codes, there was always the belief that a message once encrypted could always be unravelled. That is the golden thread that links them to the cryptologists of GCHQ today, and those yet to come.

1

TRAVELOGUE

Codes have been around for hundreds and hundreds of years but they still have a role to play in the twenty-first century.

Look at the message below. Can you decipher it? The only clue we are giving you is that these codes would not have existed before the early years of the twentieth century.

HUNGARY ITALY GERMANY EL SALVADOR.

CUBA AUSTRIA NORWAY MALTA SPAIN SPAIN THAILAND AUSTRIA THAILAND THAILAND SPAIN NORWAY.

GUYANA ANDORRA VATICAN ALBANIA.

2

DISCOVERY

Taking on the challenge of cracking codes is a voyage of discovery in itself. We have seen many puzzles in this book based on rows and columns. This time you are asked to fill the frame with the nine letters in DISCOVERY. Every letter must appear in each row and column, as well as in the nine square boxes indicated by the thicker frame lines. When completed, the shaded squares will spell out something that goes to the very heart of this book.

			D		I			
C								O
D		E				I		Y
S		O				D		E
			Y		S			
I								R
V		R	I		O	Y		D
O				E				I
E		I	V		C	O		S

3

CAREER PATH

From time immemorial, career paths for many have been fixed at birth: lord of the manor, the church, the military, the farming of food and livestock, industry in factories, working a seam in coalfields.

In this code, we follow ten twenty-first-century people in the workplace whose names are linked with a possible twenty-first-century career path. Can you work out which role they might fit in to?

1. Evie Cutex
2. Lennie Cruf
3. Abi Star
4. Theo R. Calico
5. Vera Le Popped (two words)
6. Rob Legg
7. Carlotta Ifi
8. Peter E. Runner
9. Conan Lustt
10. Julian Sort

4

BAR CODES

The word 'codes' immediately conjures up an air of mystery – espionage, dark secrets, hidden meanings. However, most twenty-first-century individuals are all too familiar with codes, in the form of bar codes, those lines and numbers which represent data that can be read by a scanner or other device, at the station or at the supermarket.

In this puzzle, all the solutions contain the word BAR, which might be at the beginning or in the middle, or towards the end of the answer. The BARRAGE of quick clues should point you in the right direction. They are all intriguing – BAR NONE!

1 Lawyer.
2 Hairdresser.
3 Make self conscious.
4 Leave a ship.
5 Plant with long, pink edible stalks.
6 Sleeveless jerkin, open at the sides.
7 Musical entertainment.
8 Sturdy waterproof cloth.
9 Sheath for a sword or dagger.
10 Small cart for use in the garden.

5

WHAT'S NEXT?

Looking to the future, it is difficult to predict what lies ahead. One thing is certain: the skills of the code-crackers will be tested as they have been for many years through history. You are challenged to see what's next in the following group.

1 CODE

OPEN

DEED

____

2 SHOUT

STEW

EATS

____

3 B

GG

N

____

6

JIGSAW

Here is a code and a jigsaw all rolled into one. A completed crossword has been broken up into NINE squares, each THREE squares by THREE. Piece the squares back together so that you can read a message going horizontally across the middle of the grid which answers the question, 'What lies ahead?'

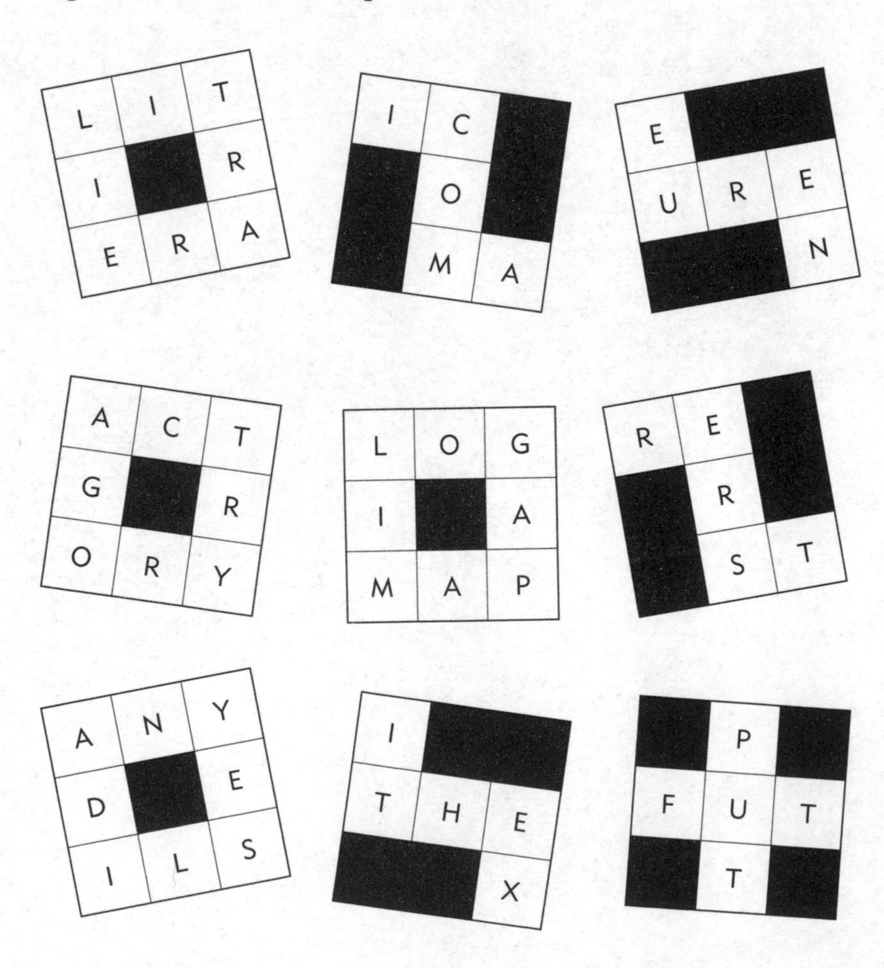

ANSWERS

CHAPTER ONE

THE CODES OF THE ANCIENTS

1

A TO Z

CODE ZEBRA. A real A to *Z* word.

1 GRAINED, **2** YARDARM, **3** XEBEC, **4** UNADORNED, **5** JACKDAW, **6** WEARILY, **7** ARCANE, **8** VEGETARIANISM, **9** NEATER, **10** MIDDLE AGED, **11** KERB, **12** ZEBRA, **13** ETERNALLY, **14** HARLEQUIN, **15** FAIRIES, **16** BIVOUAC, **17** INHERIT, **18** LASHING, **19** REFRAIN, **20** SIGNALMAN, **21** DATES, **22** OKAPI, **23** CODE, **24** QUICKTHORN, **25** TAHITI, **26** PUNDIT.

2

MAYAN MATHS

A The question mark is replaced by two dots standing for 2.

$7 + 2 - 9 = 0$

B The question mark is replaced by a horizontal line with four dots above standing for 9.

$6 \times 3 \div 9 = 2$

C The question mark is replaced by a horizontal line with three dots above standing for 8.

$8 \div 4 - 2 = 0$

D The question mark is replaced by three dots standing for 3.

$15 - 6 + 3 = 12$

3

ALPHA TO OMEGA

1 ATHENS, **2** CORINTH, **3** DELPHI,
4 SPARTA, **5** THEBES, **6** THRACE.

4

READING THE RUNES

1 ALDER, **2** HAZEL, **3** BEECH, **4** ASH, **5** WILLOW,
6 OAK, **7** ELM, **8** HOLLY, **9** SYCAMORE.

5

GLYPHS

NUMBER 6.

1 & 5 are a matching pair, as are 2 & 7, 3 & 8 and 4 & 9. That leaves 6 as the odd one out.

6

SCYTALE

The parchment is written on both sides. It reads: Beware of Greeks bearing gifts.

7

A GREEK GENIUS

The message reads: This quick brown fox vows to jump on a lazy dog.

Why is the message unusual? The most widely used letter in the English language is 'e' but this message contains all the letters of the alphabet except the letter 'e'.

CHAPTER TWO

THE CIPHER MASTERMINDS

1

GAME PLAN

Move SIX places forward.

1 TAG, **2** HOPSCOTCH, **3** LEAPFROG,
4 HIDE-AND-SEEK, **5** KITES, **6** SWING.

2

MUDDLED MATHEMATICIANS

The mathematicians have muddled up their Arabic numerals and Roman numerals. In each of the numbers 5, 6, 7, 8, 9 and 10 written out as words, a Roman numeral is hidden. In FIVE there is IV (4), in SIX there is IX (9), in SEVEN there is V (5), in EIGHT there is I (1), in NINE there is also I (1) and in TEN there are no numerals, so this is 0.

Therefore:

1 IV + V = 9

2 I + I = 2

3 0 + V = 5

4 IX + IV = 13

5 IX X 0 = 0

6 V X IX = 45

7 IV X V = 20

8 I X IV = 4

The question mark is replaced by 4.

3

CAESAR SALAD

1 CASSEROLE, **2** SAUSAGE, **3** ESCALOPE, **4** STEAK TARTARE, **5** WATERCRESS, **6** PEACH MELBA, **7** CHEESECAKE, **8** ICECREAM.

4

INVISIBLE WRITING

The thing about something invisible is that you cannot see it. Box 2 contains all the letters of the alphabet EXCEPT for A, H, I, N, S and V. The unseen letters form the word VANISH. Box 1 contains all the letters of the alphabet EXCEPT for A, D, N, O, R and U. The unseen letters form the word AROUND. Box 3 contains all the letters of the alphabet EXCEPT for D, K, S and U. The unseen letters form the word DUSK. The message reads: VANISH AROUND DUSK.

5

PIG-PEN

The message reads: There are no strangers in Freemasonry, only friends you have yet to meet.

6

MAGIC SQUARE

Each row, column and diagonal totals 65. (All the numbers added together fill 5 rows each with an equal sum, so the sum of all the numbers 1 to 25 divided by 5 gives the sum of each row, column and diagonal.)

15	16	22	3	9
8	14	20	21	2
1	7	13	19	25
24	5	6	12	18
17	23	4	10	11

7

WHEELS

1 LION (KEY LETTER T), **2** PONY (KEY LETTER I),
3 BEAR (KEY LETTER G), **4** ZEBRA (KEY LETTER E),
5 HORSE (KEY LETTER R).

The five key letters spell out TIGER.

8

WISE WORDS

GENIUS.

The key is to work out which letters appear most often. E is used twenty-four times, which is the most, so that stands for 1. S is used twenty-three times, which is the second most, so that stands for 2. N is used sixteen times and stands for 3. I is used twelve times and that stands for 4. U is used eleven times and that stands for 5. G is used ten times and that stands for 6.

9

THE VIGENÈRE SQUARE

The messages, when decrypted, read thus:

1 Missile base hidden under moor.
2 Meet contact in Vienna.
3 The major is a double agent.
4 Bomb is in the pillow.
5 Uranium on Orient Express.

10

THE CAESAR SHIFT

The messages, when decrypted, run thus:

1 The rebel Gauls have besieged us.
2 Can Brutus be trusted?
3 The hordes are at the Londinium gates.
4 Cleopatra has sealed a secret alliance.
5 Our fleet has sailed to the edge of the world.

CHAPTER THREE

THE BIBLE ENIGMAS

1

SHADOW CODE

1 DIAGRAM, **2** CARAVAN, **3** MUSTARD, **4** ATHLETE, **5** LEATHER, **6** PEBBLES, **7** THEATRE, **8** BARGAIN, **9** SEASIDE, **10** INNINGS, **11** DUNGEON, **12** COOKING, **13** MONGREL, **14** PENGUIN.

The quotation reads: I am Alpha and Omega, the beginning and the ending.

It is taken from the Book of Revelation.

2

THE WRITING ON THE WALL

1. Remove LINK, OATS, EATS, which can become BLINK, BOATS, BEATS.
2. Remove MIX and VIVID (M = 1,000, I = 1, X = 10, V = 5, D = 500.
3. Remove BEAM, CLAIM, EACH.
4. Remove ROTATOR, KAYAK, LEVEL.
5. Remove REAL and EARL.

The words left read as: WALLS HAVE EARS.

3

BEASTLY NUMBER

50130 + 51330 + 51530 = 152990

CHAPTER FOUR

PYRAMID POWER

1

PYRAMID BUILDING

Ask & Question, Collect & Accumulate, Try & Attempt, Alter & Change, Method & System, Word & Term, Unravel & Solve, Chief & Main, Discover & Find, Cipher & Code, Expression & Phrase, Labour & Work, Facts & Truth. STUDY does not have a matching word.

2

POINTED

CUBITS is the word revealed.

1 FOIL, **2** ALSO, **3** OILS, **4** CODE, **5** USED, **6** BLUE, **7** IDEA, **8** CART, **9** YEAR, **10** ITEM, **11** TREE, **12** SAVE.

3

SHIMMERING SANDS

TWELVE.

4

PATTERNS

1 M replaces the question mark. In alphabetical order, these are the letters which have vertical symmetry when written as capitals – the right-hand half mirrors the left-hand half.

2 E replaces the question mark. The pattern is made using each second letter of the twelve months of the year. A for January, E for February and so on. The final letter is the E forming the second letter of December.

CHAPTER FIVE

THE SECRET CIPHERS OF LOVE

1

LOVE'S OLD STORY

The words are the final lines of a song from the 1930s, *Love Is the Sweetest Thing*, written by bandleader Ray Noble. The lines read:

Love is the greatest thing,
The oldest yet the latest thing,
I only hope that Fate may bring
Love's story to you.

The code is created originally by writing the letters in the words above and below a line in sequence, creating two hyphenated code words in each case. Read them here by alternating between above and below the line to reveal the song words.

L V I T E R A E T H N

O E S H G E T S T I G

T E L E T E T E A E T H N

H O D S Y T H L T S T I G

I N Y O E H T A E A B I G

O L H P T A F T M Y R N

L V S T R T Y U

O E S O Y O O

2

CASANOVA'S KEYS

1 WHISKEY, **2** DONKEY, **3** MONKEY, **4** MALARKEY, **5** TURKEY, **6** KEYSTONE, **7** JOCKEYING, **8** KEY LARGO, **9** OKEY DOKEY, **10** HOKEY COKEY.

3

THE LOOK OF LOVE?

Focus on the white area and a goblet appears. Focus on the darker area and the heads of two people staring into each other's eyes appears.

4

PERSUASION ABBEY

The key to the code is to use the last letter of every sentence. The letters read: Meet by the yew tree.

5

ELGAR ARRANGEMENT

1 DEW, WED, **2** RAW, WAR, **3** DEER, REED, **4** WARD, DRAW, **5** LAGER, REGAL, **6** DRAWER, REWARD.

6

SECRET SONG

Bertie substituted the letters A, B, C, D, E, F and G for notes on the stave. However in the coded lyrics he moved each note to the note directly below on the stave. So, note A in the music is written as G in the coded lyrics. B in the music is written as A in the coded lyrics. D becomes C, E becomes D, F becomes E and G becomes F.

The message's title is ODE TO GRACE. It reads:

DEAR GRACE,
BE NOT AFRAID.
FACE GOOD AND BAD
TOGETHER FOR DECADES.

7

REBUS

REQUEST: Can I see you tonight, sweetheart? (Tin)can, eye, sea, letter U, two+knight, sweet+heart.)

REPLY: I am not for you, dear! (Letter I, joint of ham, rope knot, number 4, ewe, deer.)

CHAPTER SIX

THE AXEMAN'S ENIGMAS

1

CAVE PAINTINGS

Black circle for a body, straight tail, shaded head. There are three body shapes and colours, three tail positions and three colours for the head. Each feature appears only once in each row.

2

TUDOR SPY RING

1 WEARY, **2** ABBEY, **3** LUCKY, **4** STORY, **5** ITALY,
6 NASTY, **7** GAUDY, **8** HENRY, **9** ANGRY, **10** MARRY.

The name on the outside ring spells out WALSINGHAM, the name of Queen Elizabeth's spy master. The words which don't go in the grid make the message: Group meets where woods cross north river banks. Await there.

3

THE QUEEN'S GAMBIT

L = 17, E = 16, O = 15, A = 14, M = 13, B = 12, V = 11, G = 10,
T = 9, N = 8, Q = 7, U = 6, R = 5, I = 4, P = 3, C = 2, S = 1.

The key to solving this puzzle is to look at LEO. The letters in LEO must be 15, 16 and 17 as that is the only way you will have a total of 48 with just three letters. SCORPIO has two Os, which total 30, 32 or 34. There are five other letters. The minimum value of the five combined is 15 (1+2+3+4+5). Subtract the 15 from the total of 45 to give 30, which has to be the value of the two Os, so O has to be 15. Use this technique of looking at names with similar formations to work through all the letter values from 1 to 17.

The coded instruction reads: Urgent message. Leave at once.

4

CRYSTAL GAZING

CODE SIGN appears in the central hexagon.

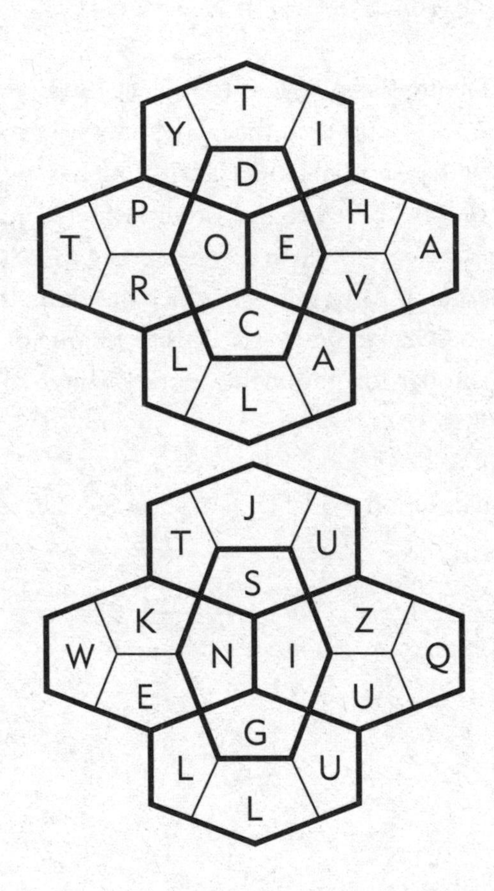

CHAPTER SEVEN

THE POETRY OF CODES

1

VERSE AND WORSE

The first letter could be A, H or R. The second letter is either O or P. The third letter is either M or T. The fourth letter is A, C or N. The fifth letter could be E, N or R. The sixth letter has to be a C. The seventh letter is either E or M. Of all the possibilities, ROMANCE is the only combination that forms a word.

2

POETRY QUEST

IDYLL is the word in the list which does not appear in the letter grid.

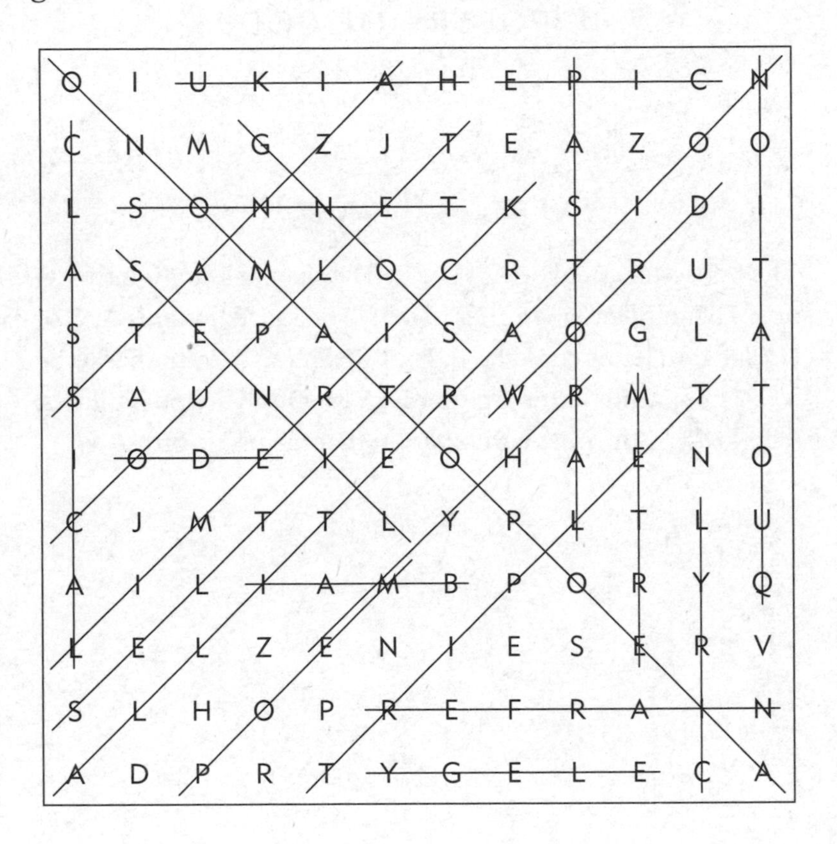

3

MOVING WORDS

Stone walls do not a prison make, nor iron bars a cage. The words are by Richard Lovelace from his seventeenth-century composition 'To Althea, From Prison'. The six columns (1 to 6) have moved so that the order became 6, 5, 3, 1, 4, 2. The seven rows (7 to 13) were also moved so that the order became 9, 12, 8, 13, 10, 7, 11.

4

BRUSH UP YOUR SHAKESPEARE

CODE A

1 Brevity is the soul of wit (*Hamlet*), **2** Parting is such sweet sorrow (*Romeo and Juliet*), **3** All the world's a stage (*As You Like It*). All the words in the quotations read backwards.

CODE B

1 Once more unto the breach, dear friends (*Henry V*), **2** The winter of our discontent (*Richard III*), **3** O brave new world (*The Tempest*). All the letters in the quotations move one letter forward in the alphabet, A becomes B, B becomes C and so on.

CODE C

1 If music be the food of love, play on (*Twelfth Night*), **2** The course of true love never did run smooth (*A Midsummer Night's Dream*), **3** I am a man more sinned against than sinning (*King Lear*). All the letters in the quotations alternate with the letters SHAKESPEARE.

5

STAR LINES

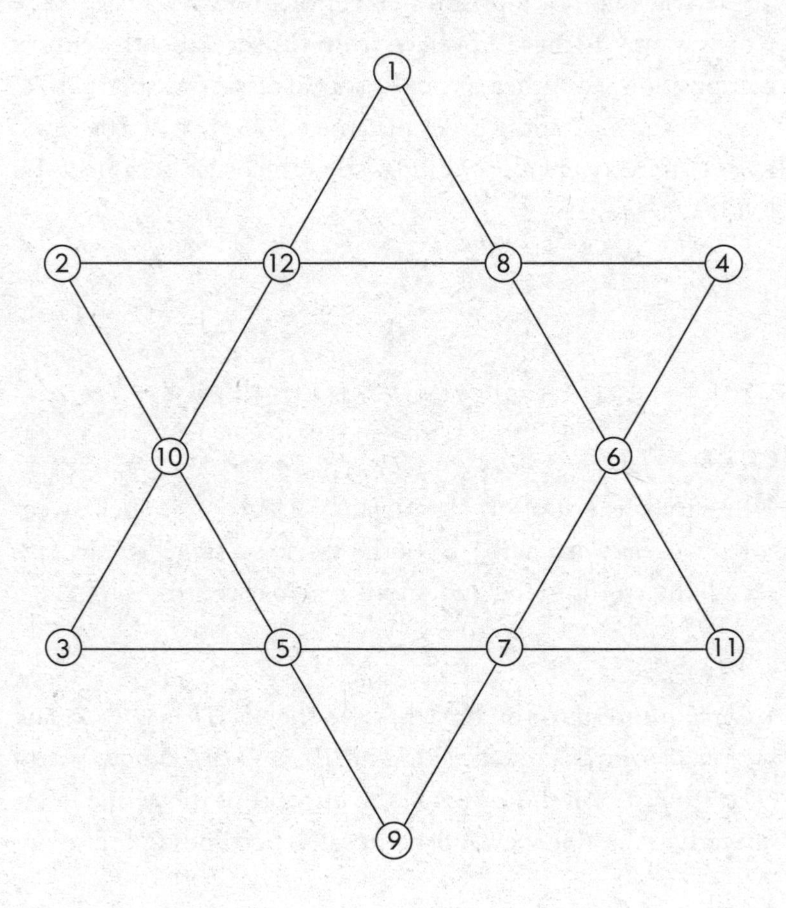

6

VALENTINE

Roses are red,
Violets are blue,
You may love me,
I don't love you.

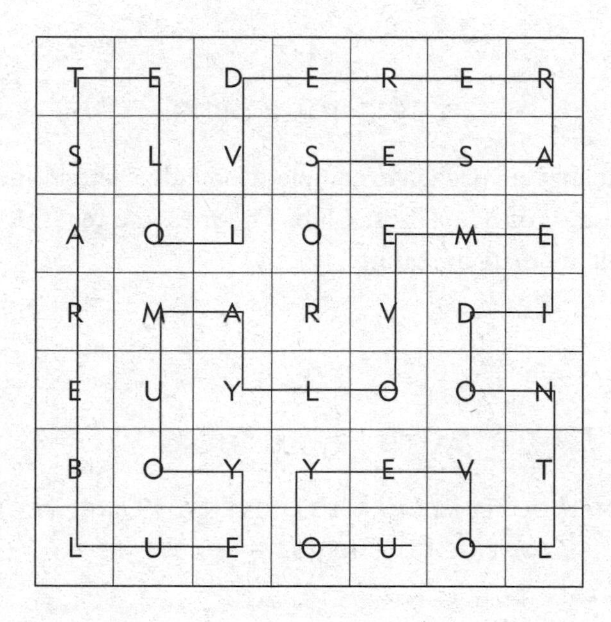

CHAPTER EIGHT

VICTORIAN ENCRYPTIONS

1

THE GHOST CLUB

Straight lines are needed to complete individual letters and solve the mystery of The Ghost Club. The message reads: Meet at ten. Tell my mate at teatime.

2

AUTHOR, AUTHOR!

1 CONAN DOYLE, **2** CHARLES DICKENS, **3** EMILY BRONTË, **4** WILKIE COLLINS, **5** THOMAS HARDY.

3

SCIENTIFIC CIPHER

Decipher the message by using the letter or letters which are symbols of the elements in the Periodic Table. The message reads: Pick up this note by noon.

P = PHOSPHOROUS, I = IODINE, C = CARBON,

K = POTASSIUM, U = URANIUM, Th = THORIUM,

S = SULPHUR, N = NITROGEN, O = OXYGEN,

Te = TELLURIUM, B = BORON, Y = YTTRIUM.

4

ANAGRAMMATIC

ACROSS: **7** Agrees, **8** Almond, **10** Aimless, **11** About, **12** East, **13** Cited, **17** China, **18** Veto, **22** Later, **23** Gardens, **24** Ardent, **25** Depart.

DOWN: **1** Cabaret, **2** Premise, **3** Seven, **4** Cleared, **5** Donor, **6** Edits, **9** Espionage, **14** Sharing, **15** Several, **16** Corsets, **19** Bleat, **20** Study, **21** Green.

5

CODED QUOTE

1 RAINCOAT, **2** UMBRELLA, **3** DOWNHILL, **4** YACHTING, **5** ANNOUNCE, **6** RECEIVER, **7** DOMINOES, **8** KOHINOOR, **9** IMMATURE, **10** PHARMACY, **11** LOVE SONG, **12** IMMORTAL, **13** NOVELIST, **14** GOVERNOR.

The name of the writer is Rudyard Kipling.

The quotation reads: A man can never have too much red wine, too many books or too much ammunition. The supplies revealed in the code therefore are: red wine, books and ammunition. Never leave home without them!

6

DECIPHERING DICKENS

1 NEWMAN NOGGS (*NICHOLAS NICKLEBY*), **2** NANCY (*OLIVER TWIST*), **3** FAGIN (*OLIVER TWIST*), **4** MARLEY (*A CHRISTMAS CAROL*), **5** TINY TIM (*A CHRISTMAS CAROL*), **6** PEGGOTTY (*DAVID COPPERFIELD*), **7** MR MICAWBER (*DAVID COPPERFIELD*), **8** SYDNEY CARTON (*A TALE OF TWO CITIES*).

7

THE DANCING MEN

The message reads: TRUST NO ONE. E is the final letter displayed. The letters in the second half of the alphabet up to U are N, O, P, Q, R, S, T and U. Given the positions of the repeated letters, trial and error leads to the only solution.

8

THE PLAYFAIR CIPHER

The decrypted messages read as follows:

1. Your contact is in Oslo.
2. Catch midnight train.
3. Formula is in painting.
4. Seize agent in Paris.
5. Meet at the border.

CHAPTER NINE

CIPHERS OF THE GREAT WAR

1

MESSAGE FROM THE TRENCHES

The key number is 678520. The last word of each sentence is an English word but when spoken out loud it sounds like a French number. Cease = *six* (6), set = *sept* (7), wheat = *huit* (8), sank = *cinq* (5), van = *vingt* (20).

2

BUILDING BRIDGES

1 COME, **2** TO, **3** TOWN, **4** HALL, **5** CROSS, **6** OVER,
7 MAIN, **8** SQUARE, **9** TAKE, **10** CARE.

Come to Town Hall. Cross over main square. Take care.

3

CODENAME

Tom is codename Wilfred. Dick is codename Squeak. Harry is codename Pip.

4

MATA HARI

RISK is the leftover word.

FILM, IDEA, LENT, MATA. CHAT, HARI, ARID, TIDE.

5

A CENSUS PUZZLE

Ada Jorkins was twenty-four and her little brother Johnnie three years of age, with thirteen brothers and sisters between. Dudeney set a linguistic/mathematical trap with the phrase 'seven times older', which could also mean 'eight times as old'

6

MOTHER AND DAUGHTER

Answer: In four and a half years – the daughter will be sixteen and a half years old and her mother forty-nine and a half years old.

7

CHANGING PLACES

There are thirty-six pairs of times when the hands exactly change places between 3pm and midnight.

8

CURIOUS NUMBERS

According to Dudeney, the three smallest numbers, in addition to 48, are 1,680, 57,120, and 1,940,448. 'It will be found,' he wrote, 'that 1,681 and 841, 57,121 and 28,561, 1,940,449 and 970, 225 are respectively the squares of 41 and 29, 239 and 169, 1,393 and 985.'

9

BEESWAX

First, the key is as follows:

1	2	3	4	5	6	7	8	9	0
A	T	Q	B	K	X	S	W	E	P

From which we then obtain:

917947476 –
408857923
―――――――
509089553

And the word BEESWAX represents the number 4,997,816

HINT: The letters represent the numbers 1, 2, 3, 4, 5, 6, 7, 8, 9 and 0. E = 9, B=4, X=6 and Q=3.

10

WRONG TO RIGHT

25938 +
25938
―――――
51876

11

THE CONSPIRATORS' CODE

```
 598 +
 507
8047
----
9152
```

12

ANOTHER POLYBIUS SQUARE

The messages, when decrypted, are as follows:

1. Under siege. Fetch fresh cohorts.
2. We meet for battle on the plain.
3. The arrows are tipped with poison.
4. Beware the maze of Knossos.
5. The Spartans are massing in the north.

CHAPTER TEN

A CONTINENT OF CODES

1

CROSSING PLACES

1 MESSINA, **2** BADAJOZ, **3** TALLINN, **4** UTRECHT, **5** LOURDES, **6** GRANADA, **7** BOLOGNA.

SALERNO is the new location in the shaded squares.

2

EURO TOUR

A ENGLAND, **B** ANDORRA, **C** DENMARK, **D** SWEDEN, **E** SPAIN, **F** ESTONIA, **G** MONACO, **H** POLAND.

3

MYSTERY IDENTITY

The hidden words all represent a letter in the NATO phonetic alphabet. This became increasingly widespread during the 1950s.

The hidden words are: **1** Lima, **2** Uniform, **3** Kilo, **4** Echo, **5** Whiskey, **6** Hotel, **7** Oscar, **8** Alpha, **9** Mike, **10** India.

The message reads: Luke, who am I?

4

FLY THE FLAG

1 Seas, e), **2** East, c), **3** Mast, d), **4** Tar, a), **5** Master, g), **6** Rear, b), **7** Steer, f), **8** Esteem, h), **9** Steamer, i), **10** Messmate, j).

5

KNEW OMISSION

Remove a letter from each word to make another word and spell out a message. The title of the puzzle becomes New Mission. The message reads: Our plan. Go east. Reach old ranch. Find red case in bushes. Papers are here. Head for Paris not Nice.

CHAPTER ELEVEN

ENIGMA AND ITS VARIATIONS

1

ENIGMATIC

The message reads: Imagine me managing an inn again.

2

IN HIDING

Each sentence contains an abbreviation for a day of the week.

1 Thurs **2** Wed **3** Fri **4** Tues **5** Mon **6** Sat

The day missing is Sun (Sunday), which will be the day of the rendezvous.

3

RECRUITING

ACROSS: **7** Charge, **9** Noodle, **10** Translator, **11** Arts, **12** Canny, **13** Tipped off, **15** Layered, **20** Laterally, **21** Error, **23** Plot, **24** Anglo-Saxon, **25** Remark, **26** Keeper.

DOWN: **1** Thermal, **2** Ironing, **3** Meals, **4** Encrypted, **5** Towards, **6** Old tiff, **8** Out like a light, **14** Balalaika, **16** Earlier, **17** Tea tray, **18** Created, **19** Hot oven, **22** Yolks.

When the grid is complete, the letters in the spaces numbered 1 to 26 spell out a message of congratulations: 'Time to contact Bletchley Park'.

4

LISTENERS

1 Peat, Taut, Hew, Too, Reed, Cymbals. Pete taught Hugh to read symbols.

2 Meat, There, Nice, Hear, Wear, Ewe, Finnish. Meet their niece here where you finish.

3 Eye, Maid, Moaner, Wring, Ate, Peels. I made Mona ring eight peals.

4 Witch, Root, Wood, Gale, Chews. Which route would Gail choose?

5 Bean, Two, Roam, Weighted, Sum, Thyme, Flue, Bye, Plain. Been to Rome. Waited some time. Flew by plane.

5

ACCESSORIES

Vera Black wore white gloves, a green hat and carried a brown handbag.

Daphne Brown wore black gloves, a white hat and carried a green handbag.

Ida Green wore brown gloves, a black hat and carried a white handbag.

Muriel White wore green gloves, a brown hat and carried a black handbag.

Daphne Brown is called in for urgent work.

6

MATHEMATICAL MINDS

3082131 + 3302131 + 4472391 = 10856653

V=0, U=1, W=2, X=3, S=4, Z=5, T=6, R=7, Y=8, Q=9.

7

TYPING POOL SLIP-UP

Instead of typing the letters on the top row of a traditional QWERTY keyboard, the typist's fingers have slipped up to the row above, i.e., the row which contains the numbers. So, Q = 1, W = 2, E = 3, R = 4, T = 5, Y = 6, U = 7, I = 8, O= 9 and P = 0.

The message reads: We require your query to Peter. Type it out. (We have added capital letters and full stops here but these were not in the original message.)

8

CROSSING THE RIVER

Although most of us are more familiar with the fox, the chicken and the sack of grain incarnation of this teaser, Carroll's version is a pleasingly dated twist! And the solution involves, for simplicity, labelling the four gentlemen and their wives thus: M1, W1; M2, W2; M3, W3; M4 and W4. And off we launch. . .

1st crossing: M1 and W1 cross; M1 returns
2nd crossing: M2 and W2 cross; M2 returns
3rd crossing: M1 and M2 cross; M2 and W2 return
4th crossing: W2 and W3 cross; M1 returns
5th crossing: M1 and M2 cross; W3 returns
6th crossing: M3 and M4 cross; M3 returns
7th crossing: M3 and W3 cross; M4 returns
8th crossing: M4 and W4 cross

9

THE FLOWER RIDDLE

This is a charming puzzle, with special appeal to fans of nature and rewilding. The answers are to be found in gardens and meadows – they are all names of wildflowers or grasses. 'Tooth of lion' is lion's tooth; then there is Ox-Eye daisy; cat's foot; foxtail grass; mouse-ear chickweed; hound's tongue; goats beard; maidenhair fern.

10

THE SUN AND THE MOON

Carroll beautifully supplied the answer via another poem:

> 'In these degenerate days,' we oft hear said,
> 'Manners are lost and chivalry is dead!'
> No wonder, since in high exalted spheres
> The same degeneracy in fact appears.
> The Moon, in social matters interfering,
> Scolded the Sun, when early in appearing;
> And the rude Sun, her gentle sex ignoring,
> Called her a fool, thus her pretensions flooring.
> ('Full' was misheard as 'fool'.)

11

CATS AND RATS

According to Lewis Carroll, the answer is both definite and yet nebulous. In terms of plain calculation, the answer would be 12. There is a big but, though: for following it through, he says, 'we find at the end of 48 minutes 96 rats are dead, and there remain 4 live rats and two minutes left to kill them in.' Given that no fractions are allowed, how would the cats arrange this final rodent bloodbath? In the original proposition, it might have been that a) six cats were needed to kill a rat, which they would do in one minute; or b) 3 cats were needed to kill a rat which they would do in two minutes; or c) 2 cats were required and the time they would need would be three minutes; or d) each cat would kill a rat by itself and take 6 minutes to do it.

The answer: if 6 cats killed six rats via methods a) or b), then the answer is still 12; but if by method c) or d), then the answer is 14.

CHAPTER TWELVE

THE STAR-SPANGLED CRYPTOGRAPHERS

1

OVAL OFFICE

1 OVAL, **2** CIRCLE, **3** SQUARE, **4** OBLONG,
5 RHOMBUS, **6** TRIANGLE, **7** PENTAGON.

All are shapes. All letters move forward six letters in the alphabet.

2

CAMBRIDGE SPY SQUARE

ANSWERS: **1** Baffled, **2** Treason, **3** College, **4** Lebanon, **5** Melinda, **6** Russian, **7** Foreign. The ship was the *Falaise*.

3

CODER

Across, left to right, top to bottom: Hawaii, Meal, Fun, Taxi, Lasso, Area, Texas, Cars, East, Cab, Wichita, USA, Bear, West, Jives, Sign, Totem, City, Hot, Cops, Hudson.

Down, left to right, top to bottom: Pasadena, Saginaw, Hills, Waco, Queens, Massachusetts, Lakes, Site, Tea, View, Cub, Sweetsop, Bronx, Zipcode, Edison, Omaha, Troy.

1 = U, 2 = S, 3 = A, 4 = E, 5 = J, 6 = M, 7 = X,

8 = C, 9 = N, 10 = Z, 11 = K, 12 = G, 13 = T, 14 = D,

15 = W, 16 = I, 17 = V, 18 = B, 19 = F, 20 = H,

21 = L, 22 = Q, 23 = Y, 24 = O, 25 = R, 26 = P.

ARLINGTON HALL is the place name.

4

EAST TO WEST

1 Indiana, **2** Maine, **3** Delaware, **4** Idaho, **5** Nevada, **6** Kansas.

The order of these states on the journey from east to west is: Maine, Delaware, Indiana, Kansas, Idaho, Nevada. NEVADA is the state, which here is the furthest south and west.

5

MORE THAN MORSE

The More Than Morse code uses triangles instead of dots and dashes. Dashes are replaced with triangles pointing downwards. Dots are replaced with triangles with their points at the top.

The patriotic words were uttered by Stephen Decatur in 1816: 'My country, right or wrong.'

6

MAN OF LETTERS

Married, Weakened, Feasted, Shadows, Thank, Hog, Went, Mow, Wet.

The quotation is: Man was made at the end of the week's work when God was tired.

The author is MARK TWAIN.

7

THE STRAIGHT AND NARROW

a) Baton Rouge (Louisana), **b)** Boston (Massachusetts), **c)** Bismarck (North Dakota), **d)** Pierre (South Dakota), **e)** Boise (Idaho), **f)** Harrisburg (Pennsylvania).

CHAPTER THIRTEEN

CODES FROM THE FUTURE

1

TRAVELOGUE

The message reads: Hi, Des. Can meet at ten. Guy and Val.

To crack the code, use the international car registration letters for the countries. Places listed alphabetically have the following codes:

ALBANIA = AL, ANDORRA = AND, AUSTRIA = A,
CUBA = C, EL SALVADOR = ES, GERMANY = D,
GUYANA = GUY, HUNGARY = H, ITALY = I,
MALTA = M, NORWAY = N, SPAIN = E,
THAILAND = T, VATICAN = V.

2

DISCOVERY

The word CODES appears in the shaded squares.

Y	R	V	D	O	I	E	S	C
C	I	S	E	V	Y	R	D	O
D	O	E	S	C	R	I	V	Y
S	Y	O	C	R	V	D	I	E
R	E	D	Y	I	S	C	O	V
I	V	C	O	D	E	S	Y	R
V	C	R	I	S	O	Y	E	D
O	S	Y	R	E	D	V	C	I
E	D	I	V	Y	C	O	R	S

3

CAREER PATH

The people's names are anagrams of the profession they may well enter into.

1 Executive, **2** Influencer, **3** Barista, **4** Chocolatier, **5** App Developer, **6** Blogger, **7** Facilitator, **8** Entrepreneur, **9** Consultant, **10** Journalist.

4

BAR CODES

1 Barrister, **2** Barber, **3** Embarrass, **4** Disembark, **5** Rhubarb, **6** Tabard, **7** Cabaret, **8** Gabardine, **9** Scabbard, **10** Wheelbarrow.

5

WHAT'S NEXT?

1 ENDS. The words make a word square that reads the same across and down. ENDS is the only option.

2 THORN. The words are all angrams of compass points. There is only one anagram of NORTH.

3 U. Letters that when read out loud sound like names of animals. Bee, Gee-gee (horse, that's a trick), Hen and Ewe.

6

JIGSAW

The words across the middle read: The Future. That certainly does lie ahead!

L	O	G	I	C		A	N	Y
I		A		O		D		E
M	A	P		M	A	I	L	S
I				P		E		
T	H	E	F	U	T	U	R	E
		X		T				N
L	I	T	R	E		A	C	T
I		R		R		G		R
E	R	A		S	T	O	R	Y

ACKNOWLEDGEMENTS

Firstly, my immense gratitude to my editor, Katie Packer, who has overseen this book with a brilliantly sharp eye, plus generosity and clarity. Equally greatly appreciated is the fantastic contribution from Feyi Oyesanya and the rest of the editorial team.

Deserving of deep admiration and awe are the cunning, diabolical and endlessly versatile skills of Sue and Roy Preston from The Puzzle House. The ingenuity of their puzzle-setting never fails to dazzle, and I am certain that if they were to hop into a time machine and whizz back to Bletchley Park, their cerebral talents would be welcomed with delight.

Huge thanks to Jill Cole and Emma Horton, who copy-edited and proofread with the dexterity of Second World War codebreakers. Another miracle-performer, with great artistic style, flair, and an insouciant way with knotty logistical problems, is the designer Dan Prescott.

Then bringing a whole range of different – though no less dazzling – skills are the marketing and PR experts Zoe Giles, Lucy Hall, Emily Patience and Joe Thomas.

Finally, my thanks to the tremendous mastermind that is my agent, Anna Power.